Geometry for the Classroom

Springer
New York
Berlin
Heidelberg
Barcelona
Budapest
Hong Kong
London
Milan
Paris
Santa Clara
Singapore
Tokyo

C. Herbert Clemens   Michael A. Clemens

Geometry for the Classroom

With 512 Illustrations

Springer

C. Herbert Clemens
Department of Mathematics
University of Utah
Salt Lake City, UT 84112
USA

Michael A. Clemens
1610 South 1900 East
Salt Lake City, UT 84108
USA

Mathematics Subject Classification: 51-01

Library of Congress Cataloging-in-Publication Data
Clemens, C. Herbert (Charles Herbert), 1939–
    Geometry for the classroom / C. Herbert Clemens, Michael A.
Clemens.
        p.   cm.
    Includes bibliographical references and index.
    ISBN 0-387-97564-0
    1. Geometry.   I. Clemens, Michael A.   II. Title.
QA453.C64   1991
516-dc20   91-14915

Printed on acid-free paper.

Camera-ready copy provided by the authors.
Printed and bound by Edwards Brothers, Inc., Ann Arbor, MI.
Printed in the United States of America.

9 8 7 6 5 4 (Corrected fourth printing, 1997)

ISBN 0-387-97564-0 Springer-Verlag New York Berlin Heidelberg
ISBN 3-540-97564-0 Springer-Verlag Berlin Heidelberg New York SPIN 10636340

Using this Book

This book is designed to give a broad preparation in elementary geometry, as well as closely related topics of a slightly more advanced nature. It is appropriate for use by high school students or as a text for preservice or in-service teachers of elementary geometry. The style of presentation and the modular format are designed to incorporate a flexible methodology for the teaching of geometry, one that can be adapted to different classroom settings. The basic strategy is to develop the few fundamental concepts of elementary geometry, first in intuitive form, and then more rigorously. The rest of the material is then built up out of these concepts through a combination of exposition and "guided discovery" in the problem sections.

The Intuition section of the book has individual sections labeled with "I." The intention here is to develop geometric intuition in order to be able to understand interesting, and perhaps rather complicated, geometric results as quickly as possible. The development is logical, but not axiomatic. We use information from arithmetic, algebra, and everyday spatial experiences without making a particular point of making these into axioms. For example, the notion of "rigid motion," that is, moving a figure around, without stretching, bending, or breaking it, is implicitly used throughout.

The Construction section ("C" pages) introduces recipes for all of the standard ruler-and-compass constructions. It is written in a manner interdependent with the Intuition section, that is, later Construction pages use earlier Intuition pages, and vice versa.

The Proof section ("P" pages) is rigorous and axiomatic, and concentrates on the properties of rigid motion. This reflects the point of view of modern geometry, that a geometry is a set, and a distinguished collection of self-transformations of that set.

Finally, there is a Computer Programs section ("CP" pages). This is included to introduce readers to the various geometric figures, constructions, and rigid motions that can be explored using the language LOGO.

Exercises are included with most sections, and are denoted with an "e." For example, the exercises for section I3 are given in I3e. Exercises include some concepts necessary for later parts of the book. So, for a complete understanding of what follows, all exercises should be done by the reader, with answers recorded in the space provided. For many routine numerical exercises, only one prototypical exercise is given. This exercise is intended as a paradigm -- in the high school classroom setting, the teacher may wish to develop a series of similar excercises as necessary to achieve student mastery. Alternatively, excercises may be taken from other texts or from companion materials developed for use with this book.

A basic year-long high school geometry course could consist in the Intuition and Construction sections, perhaps eliminating some of the more elaborate Intuition pages dealing with non-planar geometry. A more complete course, which comes to grips with the notion of rigorous mathematical proofs, would include the Proof pages. The Computer Program pages should be included as much as technological capabilities and student interest allow.

To use *Geometry for the Classroom*, readers will need a good compass, a good straightedge, and several pieces of tracing paper, as well as the appropriate pages of this book. In the classroom setting, students should enter their answers in pencil so that they can be corrected in class before final inked versions are written. These answers will then be permanently incorporated into the reader's version of the book.

About the Authors

C. Herbert Clemens is a professor of mathematics at the University of Utah. His area of mathematical research is algebraic geometry. He was the recipient of an Alfred Sloan Research Fellowship in 1974, and several National Science Foundation grants for mathematical research. He is the author of several mathematical research papers and a textbook. He was an invited speaker at the International Congress of Mathematicians in 1974, and again in 1986. He was a Parent-Teacher at Rosslyn Heights Elementary School in Salt Lake City from 1979 to 1984, during which he wrote the "Graph Paper Fractions Book." He taught the geometry course at Bryant Intermediate School during the academic year 1985-86. He has served on the Mathematical Sciences Education Board of the National Research Council and on the Committee on Mathematics of the National Research Council. He is also an editor of the Pacific Journal of Mathematics.

Michael A. Clemens attended Bryant Intermediate School from 1984 to 1986. During the academic year 1985-1986 he wrote the first draft of this book from materials presented in the geometry course taught by his father and co-author. He is currently a Ph.D. student in economics at Harvard University.

Acknowledgments

The authors would like to acknowledge the use of an Apple Macintosh Plus™ computer system in the design and formatting of the manuscript of this book. Two major applications were used; MacDraw II™ and Microsoft Word™. Within those applications, the fonts used were Courier, Symbol, Monaco, Helvetica, and Cairo by Apple Computer, and Santiago, designed at the University of Utah. The "Computer Program" section of the book was prepared using Terrapin™ Logo for the Macintosh from Terrapin, Inc.

The authors wish to thank Benjamin Clemens for his artwork and help in the use of the computer system, and the 1985-86 Geometry class at Bryant Intermediate School for their help in preparing, using, and reviewing the first draft of this book. Finally, the authors would like to thank Professor James King of the University of Washington expert advice in the preparation of the Computer Program section.

Contents

Intuition   *1*

Construction   *123*

Proof   *187*

Computer Programs   *242*

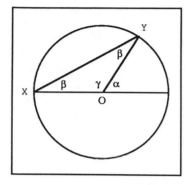

Intuition

Geometry is about shapes....

Line

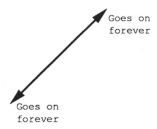

1. A line extends infinitely in two directions.

2. Given any two distinct points, there is one and only one line containing both of them.

3. Given any two points on a line, the quickest way to travel from one to the other is to stay on the line.

4. Take a point out of a line. What's left has two separate pieces:

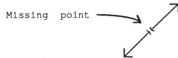

Missing point

Ray

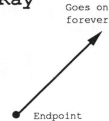

1. A ray makes up part of a line.

2. Given any two points on a ray, the quickest way to travel from one point to the other is to stay on the ray.

3. A ray has one definite **endpoint**, and extends infinitely in one direction from that point.

Segment

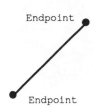

1. A segment makes up part of a line.

2. Given any two points on the segment, the quickest way to travel from one to the other is to stay on the segment.

3. Take two points, A and B, on a line. From A, send a ray down the line through B. From B, send a ray down the line through A. The points that are on both of those rays make the segment between A and B. A and B are called the **endpoints** of the segment.

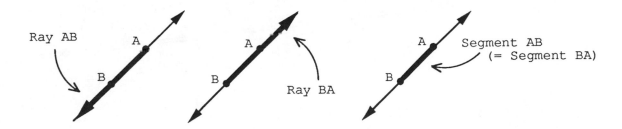

1. Is this figure a line?

2. Is this figure a line?

3. Is this figure a line?

This is
a wide
stripe.

4. Put two segments together
so that they make up one
single segment.

5. Think of a way to put
   segments together to make
   a ray. How many segments
   will you need?

6. What are the ways in
   which two <u>different</u>
   rays can intersect?

7. The distance between two points
   P and Q is the length of the
   shortest route between them.
   Suppose R lies between P and Q
   on the straight line containing
   P and Q. Tell why
   d(P,Q) = d(P,R) + d(R,Q).

   Note: "d(P,Q)" means "the distance
   between P and Q."

....and more shapes.

Triangle

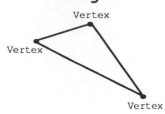

1. A **triangle** is made up of three different segments.

2. Given any one of the segments of the triangle, each endpoint of that segment is the same as one endpoint of one other segment.

3. Each segment is called a **side** of the triangle, and each endpoint of each segment is called a **vertex** (plural = vertices) of the triangle. Sides meet only at vertices.

Plane

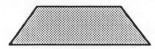

1. A plane contains many lines, in fact, an infinite number of lines.

2. Put a triangle in space. Make a line that meets two sides of the triangle in two different points. If we could draw all such lines, then, taken together, they would create a **plane**.

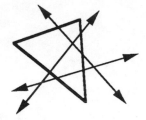
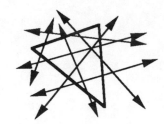
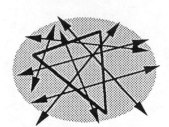

Circle

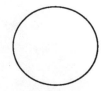

1. Pick a point in a plane, and call it "O". Pick a positive number, and call it "r".

2. A **circle** is the set of all points in the plane whose distance from O is exactly r.

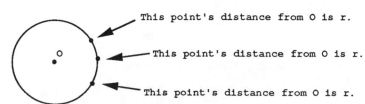

This point's distance from O is r.

This point's distance from O is r.

This point's distance from O is r.

3. r is called the **radius** of the circle, O is called the **center**.

1. Is this a triangle?
   Why or why not?

2. Do all figures made up
   of two segments lie in
   a plane? (Hint: Consider
   two cases -- 1) the
   segments intersect, or
   2) the segments don't
   intersect. If the
   segments intersect,
   make a triangle
   with one vertex at the
   intersection point and
   one side on each of the
   two segments.)

3. Can a plane intersect a
    circle in exactly two
    points? In exactly one
    point? What happens if
    the plane and the circle
    have three distinct
    points in common?

4. Can a plane intersect a
    triangle in exactly two
    points? In exactly one
    point? What happens if
    the plane and the triangle
    have three distinct
    points in common?

Polygons in the plane

Quadrilateral

1. A quadrilateral consists of four segments.

2. Given any one of the segments of the quadrilateral, each endpoint of the segment is the same as one endpoint of one other segment.

3. Each segment is called a **side** of the quadrilateral, and each endpoint of each segment is called a **vertex**. Sides meet only at vertices.

Pentagon

1. A pentagon consists of five segments.

2. Given any one of the segments of the pentagon, each endpoint of the segment is the same as one endpoint of one other segment.

3. Each segment is called a **side** of the pentagon, and each endpoint of each segment is called a **vertex**. Sides meet only at vertices.

Hexagon

1.

2.   Can you fill

                        ?

                   this in yourself?

3.

N-gon

Note that N can be any number from 3 up. The break is put in the figure because the figure will be different for different values of N.

1. An N-gon consists of N segments.

2. Given any one of the segments of the N-gon, each endpoint of the segment is the same as one endpoint of one other segment.

3. Each segment is called a **side** of the N-gon, and each endpoint of each segment is called a **vertex**. Sides meet only at vertices.

1. Explain why the figure
   below is <u>not</u> a quadrilateral.

2. Explain why the figure
   below is not a hexagon.

3. A quadrilateral can be
   divided up into triangles
   in many ways. With a
   pencil, divide the
   quadrilateral below into
   the least number
   of triangles possible.

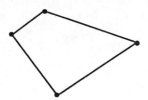

4. Divide the hexagon
   below into the least
   number of triangles
   possible.

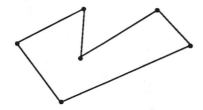

5. Imagine an N-gon. If
   N=4, what's the least
   number of triangles you
   can divide it into? If
   N=5, what's the least
   number of triangles you
   can divide it into? If
   N=6? If N=7?

   In general, an N-gon
   can be broken up into
   _____ triangles.

Angles in the plane

Angle

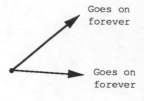

1. An angle is made up of two rays with a common endpoint.

2. The endpoint is called the **vertex** of the angle.

3. The angle divides the plane into two pieces. Whenever we have an angle, we choose one of the pieces, and call it the **inside** of the angle. The other piece is then called the **outside** of the angle.

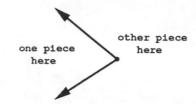

one piece here **other piece here** **We can choose to call either piece the inside of the angle.**

Congruence

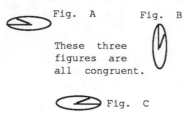

Fig. A   Fig. B

These three figures are all congruent.

Fig. C

1. We'll call two shapes congruent when you can pick one of them up and put it down exactly on top of the other one. If we want to say that figure A is congruent to figure B, we'll write A ≈ B.

2. When you're moving the first shape, to put it down on top of the second, you're not allowed to bend, stretch, or break it. But you can pick it up, flip it over, and put it back down (as many times as you want).

Size of angles

1. We say that the angle below has a measure of 180˚ (one hundred and eighty degrees):

One direction ←——•——→ Opposite direction
Vertex

Sometimes, a 180˚ angle is called a **straight angle**.

2. These two angles are congruent, and together they make a straight angle: We say that each of these angles has a measure of 90˚. Angles with a measure of 90˚ are called **right angles**.

> If two angles are congruent, they have equal measures.

> If two angles have equal measures, they are congruent.

3. Imagine 180 tiny angles, all congruent, so that when you put them all together (like we did the two right angles), you get a straight angle. We say that the measure of each tiny angle is 1˚.

1. With your ruler and
   compass, draw a segment
   on line L that is
   congruent to segment XY.

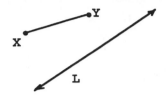

PT = The point of the compass

PL = The pencil of the compass

Use the notation
above in your
answer to describe
the placement of
your compass.

2. Three congruent angles
   with a common vertex
   together make up a
   straight angle. What
   should the measure
   of each of the angles
   be?

3. We can "add" two angles
by moving them so that
they have a common vertex
and one common ray, but
so that their insides do
not overlap. The measure
of the "sum" of two angles
is always the sum of the
measures. Here are angles
of 30°, 45°, and 90°. Use
tracing paper to build
angles of 60°, 135°, 150°,
210°, and 255°.

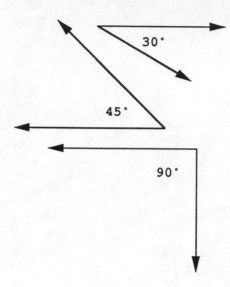

4. Use the above figures
and the tracing paper
to make angles of 15°,
and 345°

5. (SAS) Here are two
   triangles:

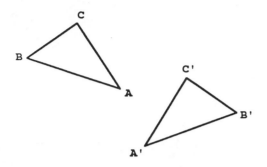

   Suppose we know:

   1) The measure of the
      angle at A is the
      same as the measure of
      the angle at A'. To say
      this, we write
      ∠A = ∠A'.

   2) The length of the side AB
      is the same as the length
      of the side A'B'. To say
      this, we write
      |AB| = |A'B'|.

   3) |AC| = |A'C'|.

   Explain why △ABC ≈ △A'B'C'.

   This fact is called the
   "Side-Angle-Side" property
   of triangles, and is
   abbreviated "SAS".

   (Hint: Pick up △A'B'C', flip
   it over, and put it down so
   that A' is on top of A and
   the direction from A' to B'
   lies on top of the direction
   from A to B. Why does B'
   then lie on top of B? Why
   does the direction from A'
   to C' lie on top of the
   direction from A to C? Why
   does C' then lie on top of
   C?)

Walking north, east, south, and west in the plane

This segment is one widget
long: •——————•

An exact copy of this widget is buried in a titanium vault, 100 miles south of Paris, France. So, whenever we need to know exactly how long one widget is, we can fly to France with our compass, travel to the vault, and set our compass to the exact length of the official widget. We will do this often throughout this book.

Much of what we do in our INTUITION pages will be based on the following assumption, which seems quite reasonable:

Stand at a point in the plane:

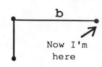

Travel north in a straight line in the plane for a distance of **a** widgets:

Now turn 90° to the right, and travel east in a straight line in the plane for a distance of **b** widgets:

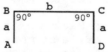

Now turn 90° to the right, and travel in a straight line in the plane for a distance of **a** widgets, directly south:

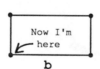

Now turn 90° to the right, and travel west in a straight line in the plane for a distance of **b** widgets:

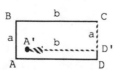

We will assume that, at the end of the trip pictured on the left, we are back to the point we started from, and that if we turned 90° to the right, and started walking again, we'd just start tracing over the picture we had before.

(This seems to be a reasonable assumption, doesn't it? Actually it is the assumption that the plane is flat and not a curved surface. To see this, try the same trip with a magic marker on the surface of a balloon.)

Said another way, if we have

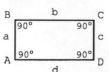

then the straight-line distance from D to A is b, and:
∠ CDA = ∠ DAB = 90°.

Suppose now we have a quadrilateral, each of whose inside angles is 90°:

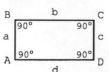

Suppose a ≠ c. Walk a units down from C. Turn right 90°, and walk b units:

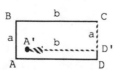

But, in the trip ABCD'A', you have to get back to where you started from.
So, A = A', and ∠D'AB = 90°.
So D'A = DA, so D' = D, and b = d.

Conclusion: Every quadrilateral, each of whose inside angles is 90°, has opposite sides of equal length.

1. Suppose on I5 that
           a = b = 1
    Then the last figure
    toward the bottom on page
    I5 is called a square
    whose side has length
    of one widget. Suppose
    it takes one can of
    paint to paint the inside
    of the square.
    Now let's take a walk. Walk
    4 widgets north, then
    6 widgets east, then 4
    widgets south, then 6
    widgets west. Would you
    get back where you started
    from? How many cans of
    paint would it take to
    paint the inside of the
    figure you traced out on
    your walk?

    Note: The number of cans
    of paint it takes to paint
    a figure is called the <u>area</u>
    of the figure (in square
    widgets).

2. Start at any point. Try
    walking 6 widgets south,
    3 widgets east, 1 widget
    north, 2 widgets west,
    five widgets north, and
    one widget west. Are you
    at the same point you
    started from? If so, how
    many cans of paint will
    it take to paint the
    inside of the figure?

3. How many cans of paint
   will it take to paint
   the checkered area?

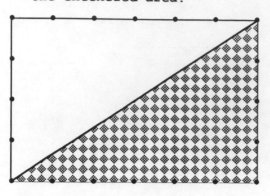

4. Find the area of the
   checkered portion:

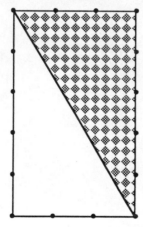

Areas of rectangles

The quadrilaterals we made in I5 and I5e are called rectangles.
(Notice that the figure in exercise #2 of I5e is not a quadrilateral)

Rectangle

We saw at the bottom of I5 that any
quadrilateral satisfying property #2
also satisfies property #3.

So, we can make one of our "I5 trips"
around any quadrilateral satisfying
property #2.

So any quadrilateral satisfying
property #2 is a rectangle.

1. A rectangle is a quadrilateral.

2. The angle at each vertex of a rectangle
   has a measure of 90°. (Extend the
   segments with endpoints at the vertex.
   That makes two rays with a common
   endpoint, so it gives an angle.
   The inside of the angle
   contains the inside
   of the rectangle.)

   Goes on
   forever

   Goes on
   forever   Inside

                                   Vertex

3. The opposite sides of a rectangle are
   of equal length.

Area of a rectangle

One can of paint is exactly
enough to paint this square
one widget on a side: 1 widget

                                    1 widget

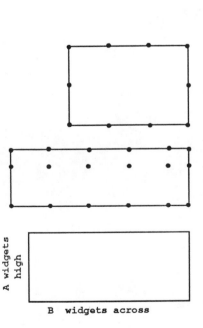

This rectangle is 3 widgets across
and two widgets high. 3 x 2 = 6,
so 6 cans of paint will paint it.
The area of this rectangle is 6
square widgets.

This rectangle is made up of 4 squares,
plus 5 half-squares, plus one quarter-
square. So its area is 6 3/4 sq. widgets.

$$(4 + {}^1/_2) \times (1 + {}^1/_2) = 4 + {}^1/_2 + 2 + {}^1/_4 = 6\ {}^3/_4$$

(Distributive Law)

A widgets high

B widgets across

Doing lots more examples we are led to
conclude that if a rectangle is A
widgets high and B widgets across, it
takes A x B cans of paint to paint it.

**Area of a rectangle =
          A x B square widgets**

1. How many cans of paint
    does it take to paint
    the larger square
    (dotted line)? Using this
    value, find how many cans
    are needed to paint the
    tilted square (shaded area).

    (Hint: Figure out how much
    paint to paint the four
    small triangles by sliding
    them together to make
    rectangles.)

Remember ●———● = one widget

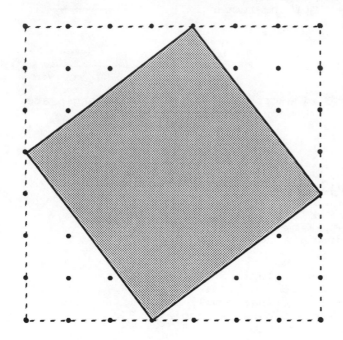

What is the area of the shaded triangle?

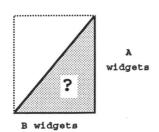

A widgets

B widgets

The shaded triangle at the left is congruent to the unshaded one, that is, the shaded triangle would fit exactly on top of the unshaded one. Together they make up a rectangle whose area is A x B sq. widgets. So:

Area of the shaded triangle

$$= \tfrac{1}{2} (A \times B) \text{ sq. widgets.}$$

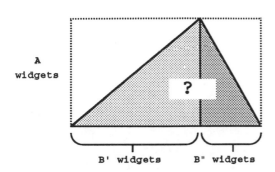

A widgets

B' widgets   **B" widgets**

Area of left-hand shaded triangle = $\tfrac{1}{2}$ (Area of left-hand rectangle) = $\tfrac{1}{2}$ (A x B') sq.w.

Area of right-hand shaded triangle = $\tfrac{1}{2}$ (Area of right-hand rectangle) = $\tfrac{1}{2}$ (A x B") sq.w.

So, the area of the big shaded triangle = $\tfrac{1}{2}$ (A x B') + $\tfrac{1}{2}$ (A x B")

$$= \tfrac{1}{2} (A \times (B' + B")) \text{ sq.w.}$$

Suppose now we want to find the area of only the darkly shaded, "tilted" triangle below:

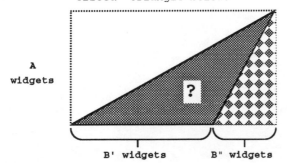

A widgets

B' widgets   **B" widgets**

Look at the big triangle made up of the shaded and checkered triangles taken together.
The area of this big triangle = $\tfrac{1}{2}$ (A x (B'+ B")) sq.w.

The area of the checkered triangle = $\tfrac{1}{2}$ (A x B") sq.w.

So, the area of the shaded triangle (the one we want) =

$$\tfrac{1}{2} (A \times (B'+ B")) - \tfrac{1}{2} (A \times B") =$$

$$\tfrac{1}{2} (A \times ((B'+ B") - B")) =$$

$$\tfrac{1}{2} (A \times B') \text{ sq.w.} \qquad \boxed{\text{Distributive Law}}$$

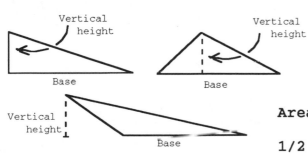

Vertical height

Base

Vertical height

Base

Vertical height

Base

Any triangle, with any side as base, is like one of the three triangles above.

Area of any triangle =

1/2 ((base)x(vertical height))

= 1/2 (b x h)

1. Slice a piece off a rectangle (starting at a corner):

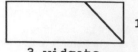

1 widget

3 widgets

Put it on the other side:

How many cans of paint does it take to paint this whole new figure?

2. The new figure we obtained in exercise #1 is called a <u>parallelogram</u>.

Now cut this new figure in half, as shown below:

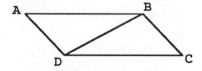

Show why segment AD is congruent to segment BC, and why segment AB is congruent to segment DC.

3. Let's go back to the
   figure we made in
   exercise #2.

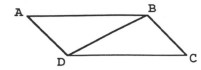

   In exercise #2 we showed
   that the three sides of
   triangle ABD were
   congruent to the three
   sides of triangle CDB.
   Later, in I10, we'll see
   that this is enough to
   conclude that triangle ABD
   is congruent to triangle
   CBD. Assume this for the
   moment, and show why the
   area of triangle CDB is
   equal to one-half the area
   of the parallelogram ABCD.

4. Using the ideas of
   exercises #1 to #3, give
   a new line of reasoning
   to show that the area
   of triangle DBC below
   is equal to $\frac{1}{2}$ (b x h).

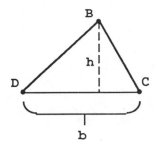

5. Look at the reasoning
    in the very first paragraph
    of I7. Notice that we take
    as "obvious" that, whenever
    we have two "angle-pieces"

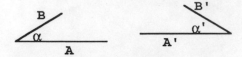

and if α = α', A = A'
and B = B', then one
"angle-piece" fits exactly
on top of the other.
Explain in your own words
why this is obvious to you
(if it is).

Adding the angles of a triangle

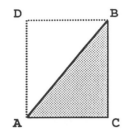

The sum of the angles in the shaded triangle ABC is 180˚.

The shaded triangle has three angles. The one at vertex A, for example, will be denoted by ∠BAC, or simply ∠A when there's no danger of confusion. (Note: The BAC part of ∠BAC means that B is on one ray of the angle, C is on the other, and A is the vertex of the angle.)

∠DBA is congruent to ∠BAC since the shaded and unshaded triangles are congruent. But, (measure ∠DBA) + (measure ∠ABC) = 90˚. So,

    (measure ∠BAC) + (measure ∠ABC) = 90˚.

So,

    (measure ∠BAC) + (measure ∠ABC) +
    (measure ∠ACB) = 90˚ + 90˚ = 180˚

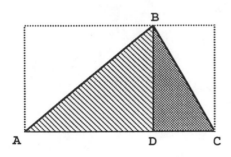

The sum of the angles in triangle ABC is 180˚.

Sum of the angles in the left-hand shaded triangle ABD equals 180˚.

Sum of the angles in the right-hand shaded triangle BDC equals 180˚.

If we add all of these angles together we get the sum of the angles in triangle ABC <u>plus</u> two extra right angles at D. So,

(sum of angles in triangle ABD) + (sum of angles in triangle BCD) = 180˚ + 180˚

(sum of angles in triangle ABC) =
        180˚ + 180˚ − (2 x 90˚) = 180˚

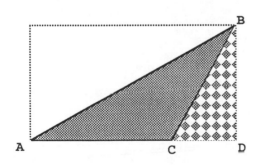

(sum of angles in triangle ABD) = 180˚

(sum of angles in triangle CBD) = 180˚

(measure ∠ACD) = 180˚

So, (sum of angles in triangle ABD) +
  (measure ∠ACD) − (sum of angles in
   triangle CBD) = 180˚.

So, (sum of angles in triangle ABC) = 180˚

The sum of the three inside angles of any triangle is $180°$.

1. What is the sum of the
   four inside (interior)
   angles of a quadrilateral?

2. What is the sum of the
   five inside (interior)
   angles of a pentagon?

3. What is the sum of the
   six interior angles of
   a hexagon?

4. What is the sum of the
   interior angles of an
   N-gon?

5. Explain why

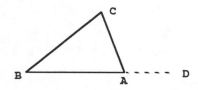

   (measure ∠DAC) =

   (measure ∠ABC) +

             (measure ∠ACB)

6. Two angles are <u>supplementary</u>
   if the they "add together"
   to make a straight angle.
   (See I4e, ex.#3.)   Can
   supplementary angles be
   congruent? If so, when?

Pythagorean theorem: $a^2 + b^2 = c^2$

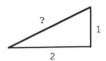

Suppose you walk two blocks east, and one block north. On this page we'll find out how far you are from where you started.

We start with a triangle whose sides have lengths a, b, and c, so that the angle opposite the side of length c is a right angle. We say γ = 90˚. So, by I8, α + β = 180˚ - 90˚ = 90˚.

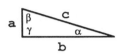

> Warning: The Greek letters stand for the <u>measures</u> of the angles shown.

Now arrange four of these triangles in the figure shown below:

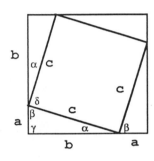

Since α + δ + β = 180˚, and α + β = 90˚, we get δ = 90˚. This means that the tilted figure (which is shaded in the right-hand copy of the picture) is a square which takes c x c = c^2 cans of paint to paint

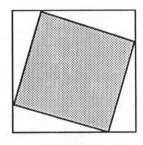

Imagine that the figure above was a room, and that the triangles were chairs. Suppose we wanted to paint the floor of the room, but didn't want to bother to paint under the chairs. We would need c^2 cans of paint.

Suppose, before we paint, we decide to rearrange the chairs as shown in the figure below. It's still going to take c^2 cans of paint to paint the floor, at least the part not covered by the chairs.

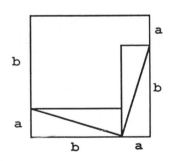

So, the shaded area on the right is still c^2. But, there's another way to figure that area. It's made up of a   b x b square and an   a x a   square. So, its area is:
$$a^2 + b^2$$
So, $a^2 + b^2 = c^2$

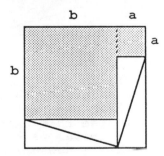

From all of this, one all-important fact emerges:
If we start with a triangle, with γ = 90˚, then

$$a^2 + b^2 = c^2$$

> Warning: The right angle <u>must</u> be the one between the sides of lengths a and b.

1. What is the formula for
    the area of the figure
    shown below?

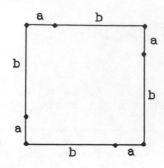

2. What is the area of this
    triangle if $\gamma = 90°$?

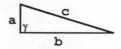

3. Use exercises #1 and #2
    to compute the formula for
    the area of the shaded
    region below:

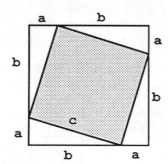

4. The shaded area in
    exercise #3 is a square
    whose side has length
    c. Use this fact, and
    exercises #1 to #3 to
    give a new way to
    conclude the
    Pythagorean theorem.

5. A triangle, one of whose
   angles measures 90°, is
   called a <u>right</u> triangle.
   The side opposite the 90°
   angle is called the
   <u>hypoteneuse</u> of the right
   triangle. (The other two
   sides are called <u>legs</u>).
   Find the length of the
   hypoteneuse in the right
   triangle shown below.

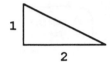

6. Find the lengths of the
   hypoteneuses in the
   right triangles shown
   below:

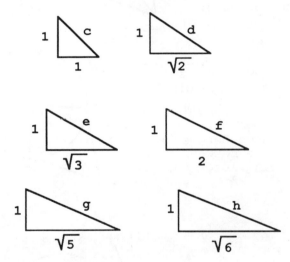

Side Side Side (SSS)

There is something we've been needing to use that we haven't talked about
yet.

The SSS Property

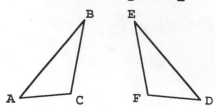

If ΔABC ("triangle" ABC) and ΔDEF are
such that: (side AB) ≈ (side DE)
              (side BC) ≈ (side EF)
              (side CA) ≈ (side FD)
then,

   ΔABC ≈ ΔDEF.

> Δ means "triangle"
> ≈ means "is congruent to"

Let's see why the SSS property is enough to
force the two triangles to be congruent:

Since DE ≈ AB, we can move ΔDEF, without
bending, stretching, or breaking it, until
D lies on top of A, and E lies on top of B.

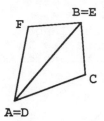

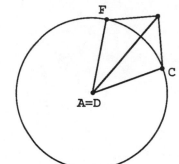

Since the distance between A and C
equals the distance between D and F,
C and F must <u>both</u> lie on the circle
at the left.

  Since the distance between B and C
  equals the distance between E and F,
  C and F must <u>both</u> lie on the circle
        at the right.

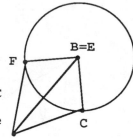

So, when A is on top of D, and B is on top
of E, both C and F must lie on each of the
two circles pictured above. But, there
are only two points common to both circles.
So, either C coincides with F at one of the
two points in which the circles intersect,
and so ΔABC and ΔDEF coincide, or C is
one of the points where the two circles
intersect and F is the other.   In this
last case, we can pick ΔDEF up by the
point F and, without bending, breaking,
or stretching it, flip it over to the
other side of the segment DE. Again, F
will come to coincide with C, and so
ΔDEF will coincide with ΔABC.

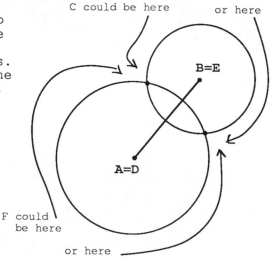

1. Show that the measure
   of ∠ ACD equals the
   measure of ∠ BCD in the
   picture below.

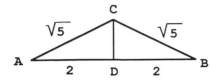

d(A,B) means "the distance between
A and B," or "the length of the
segment AB." So, above, d(A,C) =
d(B,C) = √5.

2. In the triangle below,
   all of the angles are
   equal. Why?

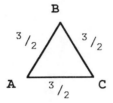

3. In exercise #2, what is
   the measure of each of the
   angles in △ABC?

4. In the triangle below,
    ∠ACB has measure 90˚.
    Give the length of
    segment AB, and show
    that ∠ CAB ≈ ∠ CBA.

> d(A,B) means "the distance between
> A and B," or "the length of the
> segment AB." So, above, d(A,C) =
> d(B,C) = 1.

5. In exercise #4 (above),
    what are the measures
    of ∠BAC and of ∠ABC?

6. The two triangles at the
    right are congruent. Pick
    out two sets of
    corresponding sides and one
    set of corresponding angles.
    Notice that "corresponding
    parts of congruent
    triangles are congruent."
    From now on we'll be using
    this principle a lot.

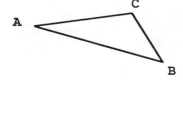

Parallel lines

Start with any rectangle:

Put an infinite number of copies of the rectangle side-by-side:

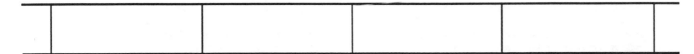

The line formed by the top of the rectangles and the line formed by the bottom of the rectangles are called **parallel** lines. Notice that parallel lines never meet.

Notation:
If lines L and M are
parallel, we write L ‖ M.

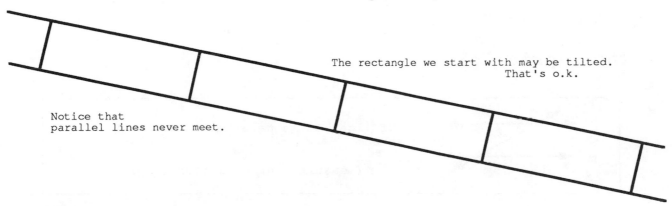

The rectangle we start with may be tilted.
That's o.k.

Notice that
parallel lines never meet.

A line L which is perpendicular to one of two parallel lines is also perpendicular to the other. (See C3.)

Let's show why this is true:

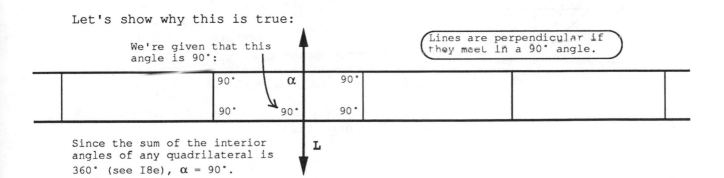

We're given that this
angle is 90°:

Lines are perpendicular if
they meet in a 90° angle.

Since the sum of the interior angles of any quadrilateral is 360° (see I8e), α = 90°.

Rectangles between parallels and the Z-principle

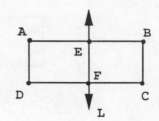

Line L is perpendicular
to the bottom side (base)
of rectangle ABCD :

We want to see that this
means that the quadrilateral
AEFD is also a rectangle.

But, as we saw at the bottom of I11, if three inside angles
of a quadrilateral each measure 90°, so does the fourth.

So, the four inside angles of AEFD are all 90°. Therefore, by
the argument we gave at the bottom of I5, |AE| = |DF|, and
|AD| = |EF|.

So, AEFD is a rectangle.

The Z-principle:

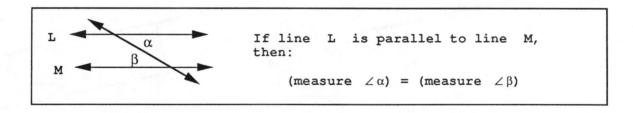

If line L is parallel to line M,
then:

(measure ∠α) = (measure ∠β)

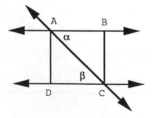

To see why the Z-principle is true, construct
DA so that the measure of ∠ CDA = 90°. By our
reasoning above, ∠ DAB = 90°. Next,
construct CB so that the measure of ∠ DCB =
90°. Again, this means that ∠ CBA = 90°.
We saw in I5 that this means that ABCD
is a rectangle. So, |AD|=|BC| and |AB|=|DC|. But,
|AC|=|AC| as well. So, by SSS (I10), ΔADC ≈ ΔCBA.
So, ∠ BAC ≈ ∠ DCA since they are corresponding
parts of congruent triangles.

1. **If line L is parallel to line M, and line M is parallel to line N, show that line L is parallel to line N. (Hint: Make the rectangles for L and M the same width as those for M and N, and pile them on top of the other ones.)**

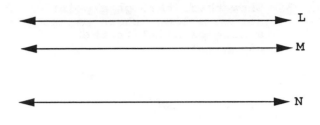

2. **We can use the Z-principle to reason to the fact that the sum of the interior angles of a triangle is 180°. Fill in the missing steps of the reasoning at the right. We will let the Greek letters stand for the _measures_ of the angles shown.**

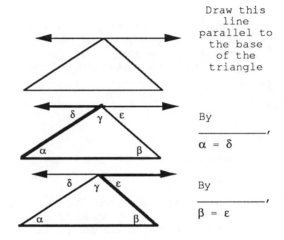

Draw this line parallel to the base of the triangle

By
_____,
α = δ

By
_____,
β = ε

Now, δ + γ + ε = 180°

So, by _____, α + β + γ = 180°

3. Show that, through a point
    not on a line, there passes
    a line parallel to the
    given line.

(Hint: Construct a
perpendicular M to L through
P, and then at P, construct
a perpendicular N to M.
Show that L and N contain
opposite sides of a
rectangle.)

4. Show that, through a point
    not on a given line, there
    passes only one line
    parallel to the given line.

(Hint: Assume that there
are two lines parallel to
L that pass through P.

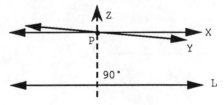

Use the Z-principle to
get a contradiction.)

5. Show the converse of the
    Z-principle:

    If the "transversal" N
    meets lines L and M
    in such a way that
    (measure ∠α) =

                  (measure ∠β),
    then lines   L   and   M
    are parallel.

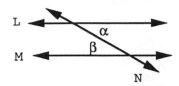

6. Show that two lines
    perpendicular to the
    same line are parallel.

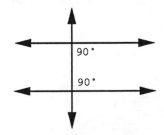

Areas: The principle of parallel slices

Suppose I'm trying to find the area
of a figure in the plane:

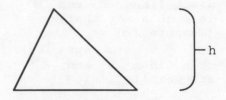

Make a lot of cuts parallel to the
base:

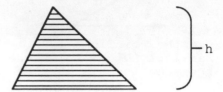

Each horizontal strip is very
close to a little rectangle.
So the area of this last
"stack of rectangles" is
very close to the area of
the original triangle above.

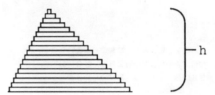

But now I can shove this
stack of rectangles against
the wall without changing
the size or shape of any
individual rectangle:

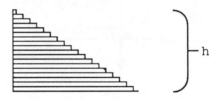

And this stack of rectangles
is very close to the triangle:

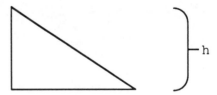

**Conclusion: When we "shove a figure against the wall,"
                    we don't change its area.**

In fact, whenever you slice a figure with parallel slices,
and shove the slices in a parallel direction, the area
stays the same.

1. When we were figuring out
    how to find the area of a
    triangle, we started with
    right triangles because
    they were the easiest.

    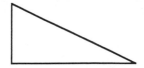

    Suppose we just know the
    formula
    $$\text{area} = \tfrac{1}{2}(b \times h)$$
    only for right triangles.
    Use the Principle of
    Parallel Slices to show that
    the same formula must be
    true for **all** triangles.

2. A quadrilateral with
    two parallel non-adjacent
    sides is called a <u>trapezoid</u>.

    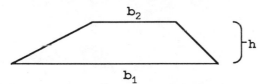

    Use the Principle of Parallel
    slices to show that the area
    of the above trapezoid is
    $$\tfrac{1}{2}(b_1 + b_2)h \ .$$

3. **Use the Principle of Parallel Slices to show that the three triangles shown below all have the same area:**

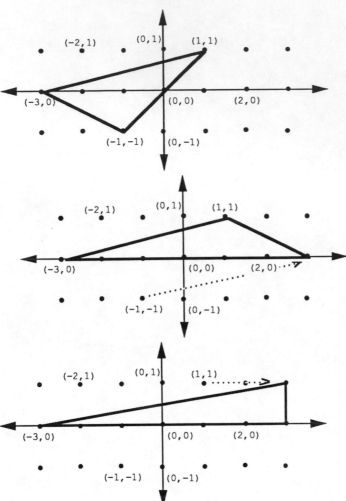

4. **Draw a trapezoid in the grid at the right, and compute its area.**

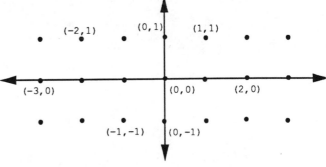

If two lines in the plane do not intersect, they are parallel

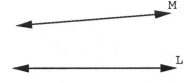

Suppose lines L and M, at right, do not meet.

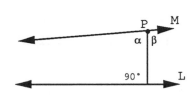

Draw a perpendicular to L through a point P
on M (see C3e).

We want to show $\alpha = \beta = 90°$, since the converse
of the Z-principle (I12e) will then let us conclude
that L and M are parallel.

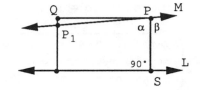

Suppose $\alpha < 90°$. We want to show that this can't
be right. Make a rectangle PQRS with a side along
L and one of the vertices at P. Let P_1 be the
point where the line M emerges again from inside
the rectangle

Now, slide the
rectangle PQRS along
the line M, and
copy it over and
over again.

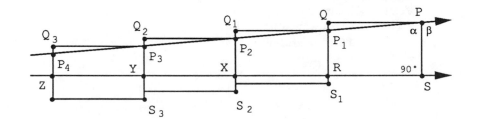

Now, lets calculate vertical distances between the points P, P_1, P_2, etc., and the
line L:

$$d(P_1,R) = d(P,S) - d(Q,P_1)$$

$$\begin{aligned}
d(P_2,X) &= d(P_1,R) - d(Q_1,P_2) \\
&= (d(P,S) - d(Q,P_1)) - d(Q_1,P_2) \qquad \text{...by substitution.} \\
&= d(P,S) - 2d(Q,P_1) \qquad \text{...since } d(Q_1,P_2) = d(Q,P_1)
\end{aligned}$$

$$\begin{aligned}
d(P_3,Y) &= d(P_2,X) - d(Q_2,P_3) \\
&= (d(P,S) - 2d(Q,P_1)) - d(Q_2,P_3) \qquad \text{...by substitution} \\
&= d(P,S) - 3d(Q,P_1) \qquad \text{...etc.}
\end{aligned}$$

So, as we got to the left with more and more rectangles, the vertical distances
between P_3, P_4, P_5, etc., and the line L, go down by the positive amount $d(Q,P_1)$ each
time. So eventually, these distances will have to go past zero, meaning that the line
M will have to meet the line L. Since M does not meet L, our assumption that
$\alpha < 90°$ must have been wrong. Going to the right instead of to the left, we see in
the same way that assuming that $\beta < 90°$ also leads to a contradiction. So the only
possibility is that $\alpha = \beta = 90°$, which means that L and M are parallel.

1. One thing we did not check
    when we were doing I14 is
    that, when we slide the
    rectangle PQRS down the
    line M, the copies we
    make as we go fit together
    side-by-side as shown in
    the picture in I14. Explain
    this by explaining why
    $\angle PP_1R \approx \angle\beta$ in the figure
    in I14.

2. Principle of Vertical Angles:

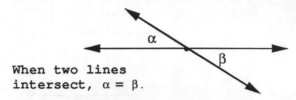

When two lines
intersect, $\alpha = \beta$.

Remember: "$\angle\alpha$" means the angle
itself, but "α" alone
means the measure of $\angle\alpha$.

Show **why** the Principle of
Vertical Angles is true.
(Hint: Find $\angle\gamma$ such that
$\alpha + \gamma = 180°$, and $\beta + \gamma$
$= 180°$.)

3. Show the Principle of
   Corresponding Angles:

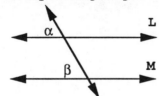

If lines L and M are
parallel, then $\alpha = \beta$.

4. **What do you think the Converse to the Principle of Corresponding Angles should be? Why is it true?**

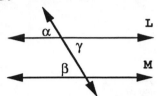

The first magnification principle: preliminary form

Suppose I have a square one widget
on a side:

    If I look at the square through a 3-power
    magnifying glass, its sides become 3
    widgets long:

What has happened to the area of our square?
Before its area was 1 sq.w. Now its area has
become 3x3x1 = 9 sq.w.

Suppose I have a square
two widgets on a side:

        If I look at the square through a
    3-power magnifying glass, its sides
                    become 6 widgets long:

What has happened to the area of our
square? Before its area was 4 sq.w.
Now its area has become 3x3x4 = 36 sq.w.

**If I magnify a square with an r-power magnifying glass,
its area is multiplied by a factor of r·r = r^2 .**

> We will use "·" to
> mean "times" when
> other notation may
> be confusing.

In the same way:

**If I magnify an interval with an r-power magnifying glass,
its length is multiplied by a factor of r.**

        r = 5

1. A square of area 14
   sq.w. is magnified
   with magnification
   factor $2\,{}^1/_2$. What
   is the area of the
   magnified square?

2. An interval of length
   7 is magnified with
   magnification factor
   $\sqrt{7}$. What is the length
   of the magnified interval?

3. A square of area 7 is
   magnified with magnification factor of the
   square root of 7. What
   is the area of the
   magnified square?

4. Suppose we have a cube
   which has edges one
   widget long:

   Suppose now we magnify
   the cube (in all
   directions) with magnification factor 2. What
   are the dimensions of
   the magnified cube?

   How many of the 1-unit
   cubes above fit into
   the magnified cube?

5. If a cube is magnified
   with magnification factor
   2, its volume is multiplied
   by _____.

   If a cube is magnified
   with magnification factor
   r, its volume is multiplied
   by _____.

The first magnification principle: final form

From exercise #5 of I15e we have a three-dimensional magnification principle:

> **If I magnify a 3-dimensional cube with an r-power magnifying glass, its volume is multiplied by r^3.**

Suppose now we have <u>any</u> 3-dimensional figure:
I claim that if I magnify it with an
r-power magnifying glass (in all directions),
the volume is multiplied by r^3.
To see this, fill up the figure (approximately)
with little cubes:

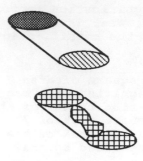

The volume of each little cube is multiplied
by a factor of r^3, so adding up all the
cubes, the volume of the magnified figure
will be multiplied by a factor of r^3:

r^3(vol. 1st little cube) + r^3(vol. 2nd little cube) + etc. + etc.
= r^3((vol. 1st little cube) + (vol. 2nd little cube) + etc.)
= r^3(volume of original 3-dimensional figure).

> Distrib-
> utive law

Using smaller and smaller cubes, I can get better and better
approximations of the original 3-dimensional figure. So:

> **If a 3-dimensional figure is magnified by a factor of r, its volume is multiplied by r^3.**

We can do the same kind of thing for 2-dimensional figures (surfaces):

> Even if the surface is curved, each little
> is approximately a square, so again by
> using the magnification principle for squares,
> and then adding up all the squares, we get:

> **If a 2-dimensional figure is magnified
> by a factor of r, its area is multiplied by r^2.**

In the same way:

> **If a 1-dimensional figure is magnified by a
> factor of r, its length is multiplied by r.**

Altogether:

If an n-dimensional figure is magnified (in all

directions) by a factor of r, its ($\begin{smallmatrix} \text{length} \\ \text{area} \\ \text{volume} \end{smallmatrix}$) is multiplied

by a factor of r^n.

1. The area of the first figure
   and the magnifying factor are
   given. Find the area of the
   second figure:

 r = 2

 r = 2 $^1/_2$

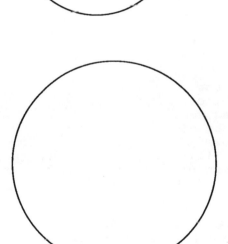

2. There is a number π
   which is the area
   inside a circle of
   radius 1. (In I17,
   we will get a pretty
   good idea of how big
   this number π is.)
   Use the Magnification
   Principle to find the
   formula for the area
   of a circle of
   radius r.

3. We will see later that
   the length of the path
   around the edge of a
   circle of radius 1 (its
   <u>circumference</u>) is 2π
   Use the Magnification
   Principle to find the
   formula for the circumference of a circle of
   radius r.

4. A sphere of radius 1 is
   the set of all points in
   3-dimensional space which
   are one unit away from
   some fixed point O.

   Later we will see that the
   surface area of a sphere of
   radius 1 widget is 4π sq.w.
   That is, it takes 4π cans of
   paint to paint the sphere.
   How many cans of paint will
   it take to paint a sphere of
   radius r widgets?

5. If the volume of the (inside
   of the) sphere of radius 1
   is $(^4/_3)\pi$, what is the volume
   of the (inside of the) sphere
   of radius r?

Area inside a circle of radius one

It takes exactly one can of paint to paint a square one widget
on a side. That is, one can of paint is exactly enough to paint
an area of one square widget (sq.w.).

This can of paint

is exactly enough
to paint this

1w.

1 w.

How much paint does it take to paint
the inside of a circle of radius
one widget?

First let's get a rough idea:

The circle of radius 1 fits inside a 2x2 box.
It takes 2·2 = 4 cans of paint to paint the
whole box so it will take less than 4 cans
to paint just the part of the box
that is inside the circle:

A box with sides $\sqrt{2}$ fits inside the circle
of radius 1: It takes $\sqrt{2} \cdot \sqrt{2}$ = 2 can of
paint to paint the inside box. So it
will take more than 2 cans of paint to
paint the entire inside of the circle.

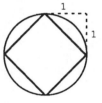

Pythag-
orean $\sqrt{2}$
Theorem

Let's let the Greek letter π stand for the number of cans of paint
it takes to paint the inside of a circle of radius one.
What we have just seen is that

$$2 < \pi < 4.$$

Let's try to get a better estimate of exactly how much paint
we need to paint the inside of a circle of radius one:

9x9 = 81 of the little squares fill up a square 1 widget
on a side. So it takes $1/81$ can of paint to paint
one of the little squares.

By counting all little squares
that lie entirely inside the
circle, we get that 212 little
squares lie entirely inside the
circle It takes 1/81 can of paint
to paint each little square. So
more than 212·(1/81) = 2.6173
cans of paint will be needed to
paint the whole inside of the
circle, since that much only paints
the little squares lying entirely
inside the circle.

By counting all little squares that
have any part inside or touching the
circle, we get that 276 little
squares entirely cover the circle and
all its inside. It takes 1/01 can of
paint to paint each little square. So
276·(1/81) = 3.4074 cans of paint will
certainly be enough to paint the inside
of the circle.

Conclusion: 2.6173 < π < 3.4074

1. The π-estimating Contest:

**Use the finest graph paper you can find to get better
lower and upper estimates for π than we did in I17.**

If you work hard at it, you should be able to conclude that
$$3.1 < \pi < 3.2.$$

Helpful hints:

1. Use the Magnification Principle—make a circle whose radius
   is, for example, 8 widgets. This will allow you to fit
   more little squares inside the circle. We know from
   exercise #2 in I16e that the area inside the circle of
   radius 8 widgets is 64π sq.w. So if, by counting little
   squares inside the circle of radius 8, we conclude
   $$a < (\text{area of circle of radius 8}) < b$$
   Then
   $$a < 64\pi < b.$$

   This inequality will then give by algebra that
   $$a/64 < \pi < b/64.$$

2. Sharpen the pencil on your compass, and draw the circle
   carefully. That will make it easier to decide which
   little squares are entirely inside the circle and which
   squares touch the circle at all.

3. Don't do a whole circle. Just do one-fourth of a circle
   and multiply your answers by 4.

Class contest: Who can get the best lower and upper estimates?
              Can anyone show that $3.13 < \pi < 3.15$?

2. C = 2πr :

Cut out several circles of radius 1 widget from the pattern at the right. Mark the center of each circle.

Let C stand for the circumference of the circle, that is, the length of a piece of string which wraps around the circle exactly once.

Cut the first circle into 8 equal pieces of pie. Arrange the pieces as shown and glue them in place. Notice that it takes π cans of paint to paint the figure at the right. Why? The length of the bumpy bottom of the figure is C/2. Why?

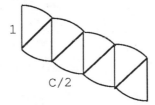

Cut the second circle into 16 equal pieces of pie. Arrange the pieces as shown and glue them in place. Notice that it takes π cans of paint to paint the figure at the right. Why? The length of the bumpy bottom of the figure is C/2. Why?

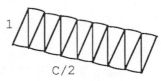

incredibly many pieces of pie

Keep going as long as you can, cutting circles into more and more pieces of pie, and arranging them in the same way. It always takes π cans of paint to paint the figure you get. Why? The length of the (less and less) bumpy bottom is always C/2. Why?

As you cut more and more finely, the figure gets closer and closer to a rectangle of base C/2 and height 1. It still takes π cans of paint to paint it. So C/2 = π. So:

    C = (circumference of circle of radius 1) = 2π

Now the circle of radius r is obtained by magnifying the circle of radius 1 with magnification factor r. Since length is one-dimensional, the Magnification Principle says that:

    C = (circumference of circle of radius r)   =   r·2π

$$= \; 2\pi r$$

When are triangles congruent?

Back in I10, we gave a rule called SSS to decide if
two triangles were congruent:

SSS: Two triangles are congruent if there is a way
of pairing off the vertices of the first with the
vertices of the second so that corresponding side
of the two triangles are congruent.

 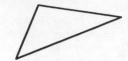

Name partner vertices A and A', B and B', C and C':

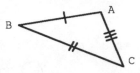 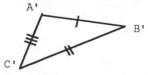

We need AB ≈ A'B'
         BC ≈ B'C'
         CA ≈ C'A'
in order to
conclude that

   ΔABC ≈ ΔA'B'C'.

But there are other useful rules too:

SAS: Two triangles are congruent if there is a way of pairing
off the vertices of the first with the vertices of the
second so that AB ≈ A'B', BC ≈ B'C', and ∠ABC ≈ ∠A'B'C'.

 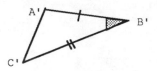

It's pretty clear that this rule is true. If you flip ΔA'B'C' over and slide it
so that B' is on top of B, and A'B' is on top of AB, then B'C' will land on
top of BC; so that forces C to land on C', and so A'C' is on top of AC.

ASA: Two triangles are congruent if there is a way of pairing
off the vertices of the first with the vertices of the
second so that AB ≈ A'B', ∠BAC ≈ ∠B'A'C',
and ∠CBA ≈ ∠C'B'A'.

It's pretty clear that this rule is true, too. If you flip ΔA'B'C' over and
slide it so that A'B' is on top of AB, then A'C' will point along AC since
corresponding angles are congruent, and B'C' will point along BC since
corresponding angles are congruent. But C is the point where the line through
AC meets the line through BC, and C' is the point where the line through A'C'
meets the line through B'C'. Since the lines are now on top of each other,
so are the points where they intersect. So C' is on top of C.

1. Suppose AB is parallel
    to DC and AD is parallel
    to BC:

    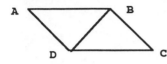

    Explain why △ABD ≈ △CDB.
    So, in particular, we
    can conclude that
    BC ≈ DA and AB ≈ CD.

2. Suppose AB is parallel
    to DC and AD is parallel
    to BC (we write AB‖DC
    and AD‖BC):

    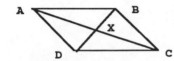

    Explain why △AXB ≈ △CXD
    So, in particular, we can
    conclude that the diagonals
    of a parallelogram bisect
    each other. (Hint: Use
    exercise #1.)

3. Show that if the diagonals
    of a quadrilateral bisect
    each other, then the quad-
    rilateral is a parallel-
    ogram. (Hint: Use the
    Principle of Vertical Angles
    and SAS, then use the
    Converse of the Z-principle
    to show that opposite sides
    of the quadrilateral are
    parallel.)

4. A triangle is called
    <u>isosceles</u> if two of
    its sides are of
    equal length.

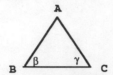

Show that, if AB ≈ AC,
then β = γ.
(Hint: Copy the
triangle, flip the
copy over, and use
SSS.)

5.

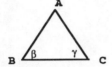

Show that, if β = γ,
then AB ≈ AC.

6. A <u>rhombus</u> is a quadrilateral
    all of whose sides are of
    equal length. Show that
    every rhombus is a parallel-
    ogram. (Hint: Draw a
    diagonal, use SSS and the
    Converse of the Z-principle.)

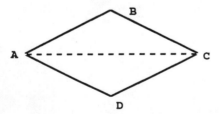

7. Show that if ABC
   and A'B'C' are
   such that
   ∠A ≈ ∠A' and
   ∠B ≈ ∠B',
   then ∠C ≈ ∠C'.

8. Show that if one can
   pair off the vertices
   of one triangle with
   the vertices of another
   so that AB ≈ A'B' and
   ∠B ≈ ∠B' and ∠C ≈ ∠C',
   then △ABC ≈ △A'B'C'.
   (We could call this
   AAS or SAA.)

9. Explain this example
   to show that ASS is
   not always true:

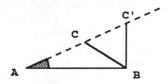

10. Show HL for right
    triangles: If two
    right triangles have
    congruent hypoteneuses
    and one leg of one is
    congruent to one leg
    of the other, then
    the two triangles are
    congruent. (Compare
    this result with
    exercise #9.)

Magnifications preserve parallelism and angles

Suppose I start at some point O and suppose
I magnify the plane in the horizontal direction by a
factor of r and in the vertical direction by a factor of r.

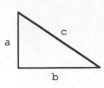

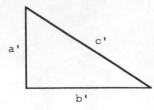

Suppose that a' = ra
and b' = rb.
Then

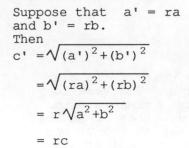

The computation on the right
(using the Pythagorean theorem)
shows that all slant distances
are also magnified by a factor
of r.

**If the plane is magnified by a factor of
r, a line L either goes to itself (if
L contains O) or goes to a line L'
which is parallel to L (if L doesn't
contain O).**

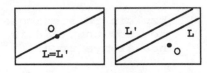

The magnification leaves O fixed. Suppose it takes P to P'.

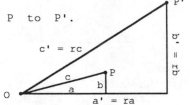

$$d(P,P') = \sqrt{(ra-a)^2 + (rb-b)^2}$$

$$= \sqrt{(r-1)^2 a^2 + (r-1)^2 b^2}$$

$$= \sqrt{(r-1)^2 (a^2 + b^2)} \ = (r-1)c$$

So,   d(O,P) + d(P,P') = c + (r-1)c = rc = d(O,P'). So P' must lie on the line
through O and P, since otherwise we would have a triangle with vertices
O, P, and P' such that the sum of the lengths of two sides of the triangle
exactly equals the length of the third. So, if L contains O, L' = L.
If L does not contain O, draw M through O parallel to L. By I11, L and
M don't meet. If L goes to L' and M goes to M' under the magnification,
then L' and M' can't meet either, since a point from L and a point
from M can't come together under magnification. So L' ∥ M' by I14. But
M' = M, so L' ∥ M. Also M ∥ L. So L' ∥ L.

**If the plane is magnified by a factor of r, angles go to
congruent angles.**

This is best seen by a picture. Suppose that, under the magnification,
L goes to L' and M goes to M'. We just saw that L is parallel to
L' and M is parallel to M':

By the Z-principle, α = β, and,
again by the Z-principle, β = α'.

So α = α'.

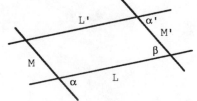

1. A 2x3 rectangle is
   magnified with
   magnification
   factor 5. How long
   is a diagonal of
   the magnified
   rectangle?

2. A square with side
   8 is magnified
   with magnification
   factor $^1/_9$. What
   is the length of
   ta diagonal of the
   "magnified" square?
   (Here the magnifi-
   cation factor is less
   than 1 so the "mag-
   nified" object is
   actually smaller than
   the original one.)
   What is the area of
   the "magnified" square?

3. One thing we really
   didn't talk much about
   in I19 is the basic
   fact that an r-power
   magnification takes
   straight lines to
   <u>straight</u> lines. Let's
   check this:

   a) Suppose P lies
   between Q and R
   on a straight
   line. Tell why
   $d(Q,P)+d(P,R)$
   $= d(Q,R)$.

   b) Suppose the
   magnification
   takes P to P',
   Q to Q', and R
   to R'. Show that
   $d(Q',P') + d(P',R')$
   $= d(Q',R')$.

   c) Tell why
   this means
   that P' is
   on the straight
   line between
   Q' and R'.

The principle of similarity

Let A, B, and C be the vertices of a triangle.
Let A', B', and C' be the vertices of another triangle.
As usual, we pair each vertex of the first triangle
with a vertex of the second so that A pairs with A',
B with B', and C with C'.

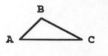 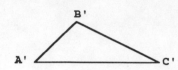   We say these two triangles are
similar if either of the two
conditions listed below are true.

<u>Condition P:</u>   Corresponding angles of the two triangles
are congruent.

<u>Condition Q:</u>   Corresponding sides of the two triangles
are proportional.

I. P implies Q.

Suppose Condition P is true.

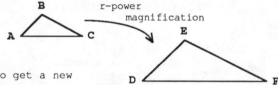

Let $r = \dfrac{d(A',B')}{d(A,B)}$

Magnify $\triangle ABC$ by a factor of r to get a new

triangle $\triangle DEF$. $d(D,E) = r \cdot d(A,B) = \dfrac{d(A',B')}{d(A,B)} \, d(A,B) = d(A',B')$.

Also $\angle D \approx \angle A$ since magnification preserves angles by I19.

Furthermore, $\angle A \approx \angle A'$ by Condition P. So $\angle D \approx \angle A'$.

By the same line of reasoning, $\angle E \approx \angle B'$.

So, by ASA, $\triangle DEF \approx \triangle A'B'C'$. So $d(E,F) = d(B',C')$.

But $d(E,F) = r \cdot d(B,C)$ since EF comes from magnifying BC by a factor of r.

Therefore $\dfrac{d(B',C')}{d(B,C)} = \dfrac{d(E,F)}{d(B,C)} = \dfrac{r \cdot d(B,C)}{d(B,C)} = r$.

By the same line of reasoning, $\dfrac{d(C',A')}{d(C,A)} = r$. So Condition Q is true.

II. Q implies P.

Suppose Condition Q is true. Then $\dfrac{d(A',B')}{d(A,B)} = \dfrac{d(B',C')}{d(B,C)} = \dfrac{d(C',A')}{d(C,A)} = r$.

On the other hand, we can magnify $\triangle ABC$ by a factor of r to get a
triangle $\triangle DEF$.

But $d(D,E) = r \cdot d(A,B) = \dfrac{d(A',B')}{d(A,B)} \, d(A,B) = d(A',B')$

$d(E,F) = r \cdot d(B,C) = \dfrac{d(B',C')}{d(B,C)} \, d(B,C) = d(B',C')$

$d(F,D) = r \cdot d(C,A) = \dfrac{d(C',A')}{d(C,A)} \, d(C,A) = d(C',A')$

So, by SSS, $\triangle DEF \approx \triangle A'B'C'$. This means that $\angle D \approx \angle A'$.

But $\angle D \approx \angle A$ since magnification preserves angles. So $\angle A' \approx \angle A$.

By the same line of reasoning, $\angle B' \approx \angle B$ and $\angle C' \approx \angle C$.
So Condition P is true.

1. If BC is parallel to
   DE, show why △ABC
   is similar to △ADE.

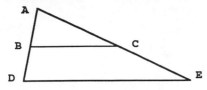

   (From now on we will
   write "~" to mean
   "is similar to." So
   we can rewrite this
   problem as follows:
   Show that, if BC ‖DE,
   then △ABC ~ △ADE.)

2. For a figure as in
   exercise #1, suppose
   d(A,B)=2, d(A,C)=3,
   and d(A,D)=5. Find
   d(A,E).

3. For a figure as in
   exercise #1, suppose
   d(A,C)=2, d(A,D)=5,
   and d(A,E)=10. Find
   d(A,B).

4. For a figure as in
   exercise #1, suppose
   d(A,C)=2, d(B,C)=5,
   and d(A,E)=10. Find
   d(D,E).

5.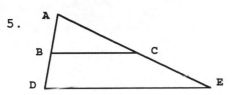

   Suppose △ABC ~ △ADE.
   Show that BC ‖DE.

6. To show that two
   triangles are sim-
   ilar, it's enough
   to show that <u>two</u>
   of the angles of
   the first are con-
   gruent to corres-
   ponding angles of
   the second. Why
   is that?

7.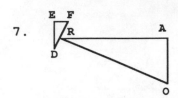

   Suppose OR ⊥ DF, and
   that all other seg-
   ments are horizontal
   or vertical. Show
   that △AOR ~ △EFD.

   > "⊥" means
   > "is perpen-
   > dicular to."

8. <u>Condition R:</u>
   Two sides of one triangle
   are proportional to two
   sides of another, and the
   angles included by the
   two sides are equal.

   Show: R implies P.
   (Hint: The proof is very
   similar to "Q implies P,"
   but use SAS instead of SSS.)

9. Show: P implies R.

10.

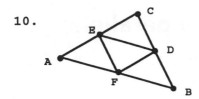

D is the midpoint of
the segment BC, E is
the midpoint of AC,
and F is the midpoint
of AB. What can you
say about the segments
and triangles in this
picture? Give reasons
for your answers.

11.

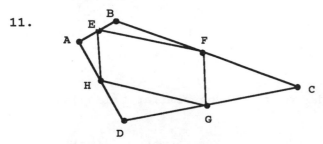

The quadrilateral EFGH has
as its vertices the midpoints
of the sides of the quad-
rilateral ABCD. What can you
say about the quadrilateral
EFGH? (Hint: You may want
to draw in AC momentarily
to help visualize things.)

12.   Suppose ΔAEB is a right
      angle and AB ⊥ CE. Show
      that the triangles AEB,
      ACE, and ECB are all
      similar.

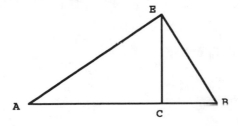

Proportionality of segments cut by parallels

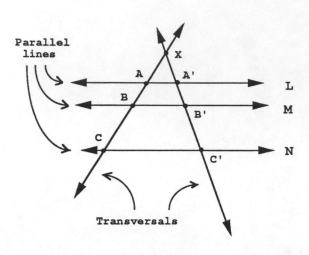

Given: Lines L, M, and N
are parallel.

Show: ΔXAA' ~ ΔXBB' ~ ΔXCC'

Proof: ∠XAA' ≈ ∠XBB'   Principle of
∠XA'A ≈ ∠XB'B   Corresponding
                      Angles (I14e)

ΔXAA' ~ ΔXBB'   I20.

ΔXBB' ~ ΔXCC'   Same argument

So, by I20, corresponding sides of the three similar triangles
in the above picture are proportional:

> To save space
> we'll write
> |AB| for d(A,B)
> etc.

$$\frac{|XB|}{|XA|} = \frac{|XB'|}{|XA'|} \quad \text{and} \quad \frac{|XC|}{|XA|} = \frac{|XC'|}{|XA'|} \quad .$$

Now let's use these two equations and do some algebra:

1) $\quad \dfrac{|XC|}{|XA|} - \dfrac{|XB|}{|XA|} = \dfrac{|XC'|}{|XA'|} - \dfrac{|XB'|}{|XA'|} \quad \text{so} \quad \dfrac{|BC|}{|XA|} = \dfrac{|B'C'|}{|XA'|}$

2) $\quad \dfrac{|XB|}{|XA|} - 1 = \dfrac{|XB'|}{|XA'|} - 1$

3) $\quad \dfrac{|XB|}{|XA|} - \dfrac{|XA|}{|XA|} = \dfrac{|XB'|}{|XA'|} - \dfrac{|XA'|}{|XA'|} \quad \text{so} \quad \dfrac{|AB|}{|XA|} = \dfrac{|A'B'|}{|XA'|}$

4) $\quad \dfrac{|BC|}{|B'C'|} \underset{1)}{=} \dfrac{|XA|}{|XA'|} \underset{3)}{=} \dfrac{|AB|}{|A'B'|}$

5) $\quad \dfrac{|BC|}{|AB|} \underset{4)}{=} \dfrac{|B'C'|}{|A'B'|}$

**Three parallel lines cut transversal lines
in proportional segments.**

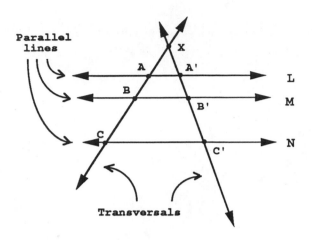

1. Fill in the missing value:

|AB|	|BC|	|A'B'|	|B'C'|
5	6	7	__
2	__	6	9
__	7	4	7
1	5	__	15

2. Fill in the missing steps:

                                             Reason

    a) $\dfrac{|BC|}{|AB|} = \dfrac{|B'C'|}{|A'B'|}$ _____

    b) $\dfrac{|BC|}{|AB|} + 1 = \dfrac{|B'C'|}{|A'B'|} + 1$ _____

    c)   _____

    d) $\dfrac{|BC| + |AB|}{|AB|} = \dfrac{|B'C'| + |A'B'|}{|A'B'|}$ _____

    e) $\dfrac{|AC|}{|AB|} = \dfrac{|A'C'|}{|A'B'|}$ _____

3. Why does $\dfrac{|XA|}{|XB|} = \dfrac{|AA'|}{|BB'|}$?

4. Notice that in the I21
   diagram, there is nothing
   to prevent the line L
   from passing through the
   point X. Use this to
   explain why the following
   theorem is true:

   <u>Theorem</u>: A line parallel
   to the third side of a
   triangle cuts the other
   two sides into propor-
   tional segments.

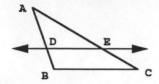

5. Show the converse to
   the theorem in exercise
   #4, that is:
   <u>Theorem</u>: Suppose

   $$\frac{|DB|}{|AD|} = \frac{|EC|}{|AE|} .$$

   Then DE is parallel to
   the line through BC.

   (Hint: Show that

   $$\frac{|AB|}{|AD|} = \frac{|AC|}{|AE|}$$

   and then use exercises
   #8 and #5 of I20e.)

6.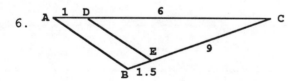

   Why is AB‖DE?

7. In exercise #6, find $\frac{|DE|}{|AB|}$.

8. Is there enough information
   in exercise #6 to determine
   |DE|?

9. In the coming pages, we will start
    with any triangle and construct
    special lines associated to that
    triangle, and we will show that
    these lines have surprising
    special properties. For example,
    we will see in I23 that the three
    altitudes of any triangle pass
    through a common point. (See C16
    for definition of altitude.) Also
    we will see in C21 that the
    perpendicular bisectors of the
    three sides of any triangle pass
    through a common point. The
    following exercise shows that
    these two facts are related.

    Construct your favorite triangle
    and construct its three altitudes:

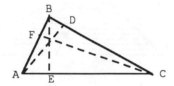

    Now construct a perpendicular to
    AD through A, a perpendicular to
    BE through B, and a
    perpendicular to CF through C:

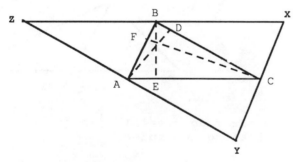

    Show that ACBZ is a parallelogram,
    and that ACXB is a parallelogram.

    Conclude that ZB ≈ BX, so that BE is
    the perpendicular bisector of ZX.

    Conclude in the same way that CF is
    the perpendicular bisector of XY,
    and AD is the perpendicular bisector
    of XZ.

    So the perpendicular bisectors of
    the sides of △XYZ are the same as
    the altitudes of △ABC.

Finding the center of a triangle

By the **center** of a triangle, we mean the point inside the
triangle where it would balance if it is put on the tip of
one's finger:

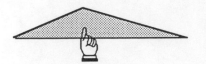

(You have to imagine that the
triangle is made out of a thin
sheet of steel which doesn't bend!)

Let's try to figure out how
to find the center of a triangle:

Suppose we cut the triangle
into horizontal slices. If
we were to take out one slice
and try to balance it, we would
put our finger halfway along
the slice, that is, at the
midpoint of the slice. Mark
the midpoint of each slice
and reassemble the triangle:

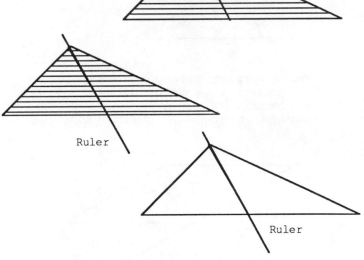

Suppose I put a thin, weight-
less ruler along the line
formed by all the midpoints.
Since each individual slice
of the triangle balances
along this line, the entire
triangle will balance along
the ruler:

**So to balance the whole triangle on our finger, we'll
have to put our finger at some point along the ruler.**

But we could repeat the same
reasoning starting from one
of the other sides of the
triangle:

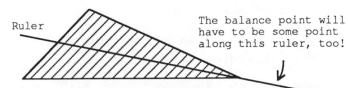

The balance point will
have to be some point
along this ruler, too!

So here's the center:

(It didn't matter which
two sides we worked from
to find the center.)

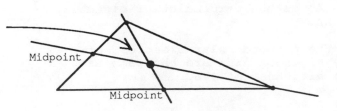

1. To find the midpoint of
   a segment in the (x,y)-
   plane:

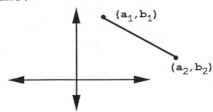

$$midpoint = (\ \frac{a_1+a_2}{2}, \ \frac{b_1+b_2}{2}\ )$$

x-coordinate   y-coordinate
  is average   is average
of x-coords.   of y-coords.
of endpoints   of endpoints

Find the coordinates of
the midpoints of all the sides
of the polygon at the right.

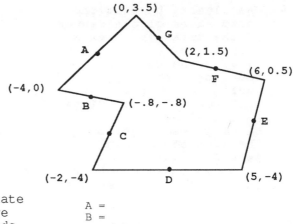

A =
B =
C =
D =
E =
F =
G =

2. In the next three problems, we
   will find the coordinates of the
   center of the triangle below:

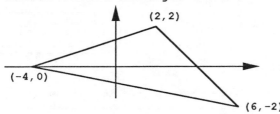

First find the coordinates of
the midpoints of the three
sides, and plot these points

3. A <u>median</u> of a triangle is a segment
   between a vertex of the triangle
   and the midpoint of the opposite
   side. Draw two of the medians
   of the triangle in exercise #2.
   By I22, they meet in the center
   of the triangle. Since every
   median contains the center of
   the triangle, the third median
   had better pass through the point
   of intersection of the other two.
   Check this by drawing the third
   median.

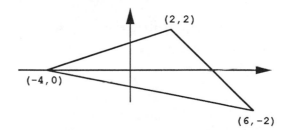

4. Use algebra to find the coordinates of the center of the triangle in ex.#2. Steps:
a) Find the equation of the line containing a median, using the fact that you know two points on the line.
b) Find the equation of the line containing a second median.
c) Solve the two linear equations simultaneously.

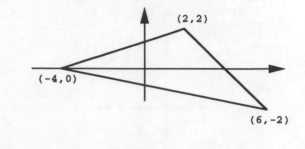

5. <u>Concurrence Theorem for Medians</u>:
The three medians of any triangle pass through a common point.

The reasoning of I22 showed why this theorem is true. In this exercise and the next is another argument. Fill in the missing steps.

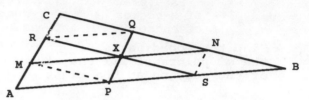

Construct a line parallel to AC and passing through X. Let P and Q denote the points at which the line meets the other two sides of the triangle. In the same way, contruct RS||CB and MN||AB.

Given: AB||MN, BC||SR, AC||PQ, X is midpoint of MN and of SR.
Show: X is also midpoint of PQ.

> The "Given" is true when X is on the medians from vertices A and C.

	Reason
Proof: ΔRXM ≈ ΔSXN	**SAS**
∠XRM ≈ ∠XSN	- - - - - - - - - - - - - - - - - - - -
MR‖SN	- - - - - - - - - - - - - - - - - - - -
XSNQ is a parallelogram	Definition of parallelogram
XQ ≈ SN	Exercise #1 of I18e
PX‖SN	PX‖MR and MR‖SN
XNSP is a parallelogram	- - - - - - - - - - - - - - - - - - - -
PX ≈ SN	- - - - - - - - - - - - - - - - - - - -
PX ≈ XQ	- - - - - - - - - - - - - - - - - - - -

6.   Given: PQ||AC and PX ≈ XQ

Show: AY ≈ YC

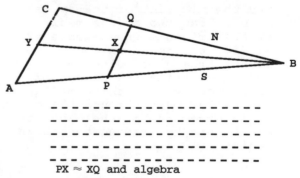

Proof: ΔBXQ ~ ΔBYC
|XQ|/|YC| = |BX|/|BY|
ΔBPX ~ ΔBAY
|PX|/|AY| = |BX|/|BY|
|PX|/|AY| = |XQ|/|YC|
|AY| = |YC|

- -
- -
- -
- -
- -
PX ≈ XQ and algebra

Concurrence theorem for altitudes of a triangle

Theorem: **The three altitudes of any triangle pass through a common point.**

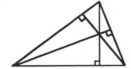

"¬"
means
right
angle

We will see why this theorem is true <u>indirectly</u>. Namely, we will suppose there is some "bad" triangle for which the theorem is not true—then we will show that that triangle would in fact have some property that contradicts something that we have already shown to be true. So the only possibility is that the "bad" triangle doesn't exist. So the theorem must be true for all triangles.

Suppose we have a "bad" triangle ABC: Then the perpendicular GX does not pass through C. Let C' be the point in which the line GX meets the line AC. Let C" be the point in which the line GX meets the line BC. Then C' ≠ C".

∠ AXF ≈ ∠ BXE   Vertical angles

ΔAXF ~ ΔBXE   Ex.#6 of I20e

By the same reasoning:

ΔAXG ~ ΔC"XE   and   ΔBXG ~ ΔC'XF

So by Condition Q of I20:

$$\frac{|AX|}{|BX|} = \frac{|XF|}{|XE|} \qquad \text{so} \quad |AX|\cdot|XE| = |BX|\cdot|XF| \qquad (1)$$

$$\frac{|AX|}{|C"X|} = \frac{|XG|}{|XE|} \qquad \text{so} \quad |C"X| = |AX|\cdot|XE|/|XG| \qquad (2)$$

$$\frac{|BX|}{|C'X|} = \frac{|XG|}{|XF|} \qquad \text{so} \quad |C'X| = |BX|\cdot|XF|/|XG| \qquad (3)$$

Now by substituting equation (1) in equation (3): |C'X| = |AX|·|XE|/|XG|

Combining this last equation with equation (2), we get: |C'X| = |C"X|

But now let's look again at our "bad" triangle ABC:

If |C'X| = |C"X|, then C' = C", since C' and C" lie on the same line in the same direction from X. But this contradicts the fact that, for "bad" triangles, C' ≠ C" (look back at the beginning of this proof). Since we got to a contradiction by assuming that there was a "bad" triangle for which the theorem wasn't true, it must be that the assumption itself was wrong— namely, there are <u>no</u> triangles for which the theorem is not true!

1. An angle is called <u>obtuse</u>
   if its measure is more
   than 90°. If one of the
   angles of a triangle is
   obtuse, two of its altitudes
   will lie outside the triangle.
   Use exercise #1 of C3e to
   construct the three
   altitudes of the triangle
   at the right. Extend them
   to find the point where
   they all meet.

2. Where do all the altitudes
   of a right triangle meet?

3. An angle of a triangle is
   <u>acute</u> if its measure is
   less than 90°. If all three
   angles of a triangle are
   acute, where do its altitudes
   meet?

4. Using your straight edge and
   compass and C3e, construct
   all the medians and all the
   altitudes of the triangles
   shown. Mark the concurrence
   points for each.

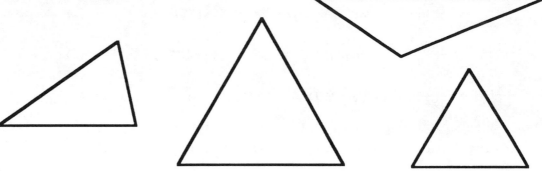

There are two other concurrence theorems, one for angle bisectors of a triangle,
the other for perpendicular bisectors of sides of a triangle. Since they have
alot to do with inscribing circles in triangles and circumscribing circles around
triangles, we will discuss them in pages C20 and C21.

5. In the diagram below, show
    why:

    (area ΔAXC)/(area ΔAXB)

    = |CD|/|DB|

    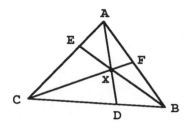

    (Hint: First show
    (area ΔADC)/(area ΔADB).
    = |CD|/|DB|.
    Then show:
    (area ΔXDC)/(area ΔXDB)
    = |CD|/|DB|.)

6. Ceva's Theorem states that,
    in the diagram in exercise
    #5, the lines AD, BE, and CF
    are concurrent (pass through
    a single point) at X inside
    the triangle if and only if:

    $$\frac{|AE|}{|EC|} \cdot \frac{|CD|}{|DB|} \cdot \frac{|BF|}{|FA|} = 1$$

    Show why Ceva's Theorem
    is true. (Hint: use the
    principle you discovered
    in #5.)

7. Use exercise #5 to show
    why the Concurrence
    Theorem for Medians is
    true.

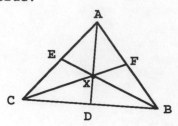

8. Suppose all three
    altitudes of a triangle
    lie inside the
    triangle. Use Ceva's
    Theorem to show why the
    Concurrence Theorem for
    Altitudes is true.

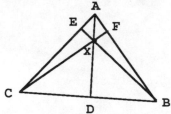

    (Hint: You will need to
    use the converse; that is,
    the direction of Ceva's
    Theorem which says that

    if: $\dfrac{|AE|}{|EC|} \cdot \dfrac{|CD|}{|DB|} \cdot \dfrac{|BF|}{|FA|} = 1$

    <u>then</u> the lines AD, BE,
    and CF are concurrent.)

Inscribing angles in circles

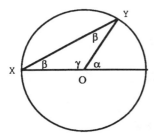

Since OX and OY are radii, |OX| = |OY|. So, by exercise #4 of I18e, the angles OXY and OYX have the same measure.

So β + β + γ = 180°.

But α + γ = 180°.

So   2β = α .

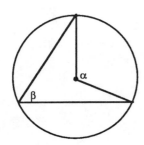

Here again   2 β = α.

To see this, just draw in a diameter of the circle:

$$2\beta_1 = \alpha_1$$

$$2\beta_2 = \alpha_2$$

so $2(\beta_1 + \beta_2) = (\alpha_1 + \alpha_2)$

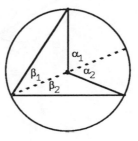

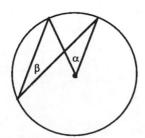

Here again   2 β = α.

To see this, just draw in a diameter of the circle:

$$2\beta_1 = \alpha_1$$

$$2\beta_2 = \alpha_2$$

so $2(\beta_1 - \beta_2) = (\alpha_1 - \alpha_2)$

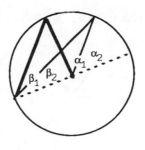

The measure of an angle inscribed in a circle is always one-half of the measure of the central angle which cuts out the same arc on the circle.

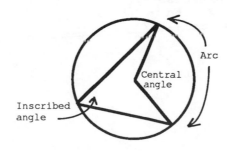

1. The angle α is incribed
   in a semi-circle. What
   is its measure?

2. Explain why
   ΔAXB ~ ΔCXD.

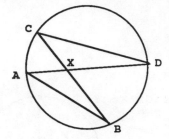

3. A segment whose endpoints
   lie on a circle is called
   a <u>chord</u> of the circle.
   Suppose chords AD and BC
   meet in a point X. Show
   that

   $$|AX| \cdot |XD| = |CX| \cdot |XB|.$$

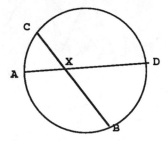

4. Suppose, in exercise #3,
   |AX|=2, |XD|=4, and
   |CX|=3. What is |XB|?

5. Show that △PXQ' ~ △QXP'.

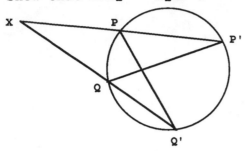

(Hint: To show that
∠XPQ' ≈ ∠XQP',
show first that
∠Q'PP' ≈ ∠P'QQ'.)

6. Show that

$$|XP| \cdot |XP'| = |XQ| \cdot |XQ'|.$$

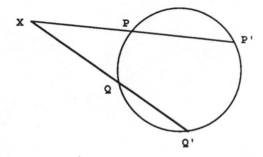

7. Suppose, in exercise #6,
|XP|=6, |XP'|=12, and
|XQ'|=10. What is |XQ|?

8. Let's use $\overset{\frown}{AB}$ to mean
   the arc between A and
   B and
   $|\overset{\frown}{AB}|$
   to mean the measure
   of the central angle
   given by the arc AB:

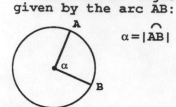

$\alpha = |\overset{\frown}{AB}|$

   Show that, in the
   picture below,
   $\alpha = \frac{1}{2}|\overset{\frown}{AB}| + \frac{1}{2}|\overset{\frown}{CD}|$.

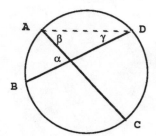

   (Hint: $\alpha = \beta + \gamma$ by
   exercise #5 of I8e)

9. Show that, in the
   picture below,
   $\alpha = \frac{1}{2}|\overset{\frown}{AB}| - \frac{1}{2}|\overset{\frown}{CD}|$.

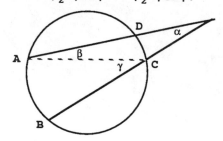

10. Suppose, in exercise #9,
    $|\overset{\frown}{CD}| = 20°$, and $\alpha = 10°$.
    What is $|\overset{\frown}{AB}|$?

Fun facts about circles, and limiting cases

The facts we have been obtaining in the exercises of I24e are really important, so let's make a list:

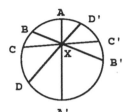

$$|AX| \cdot |XA'| = |BX| \cdot |XB'| = |CX| \cdot |XC'| = |DX| \cdot |XD'|$$

The product of the lengths of the two segments of a chord through X depends only on the position of X. It does not depend on which chord through X we take. (Exercise #3 of I24e.)

$$|XA| \cdot |XA'| = |XB| \cdot |XB'| = |XC| \cdot |XC'| = |XD| \cdot |XD'|$$

The product of the distances from a point X (outside a circle) to the two points in which a line through X meets a circle depends only on the position of the point X. It does not depend on which line through X we take. (Exercise #6 of I24e.)

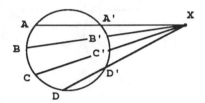

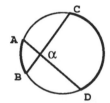

The measure of an angle α formed by two chords is the average of the central angles of the arc CD cut by ∠α and the arc AB cut by the vertical angle to ∠α. (Exercise #8 of I24e.)

The measure of an angle α formed by two lines through a point X (outside a circle) is one-half the difference of the central angles of the two arcs cut by the angle. (Exercise #9 of I24e.)

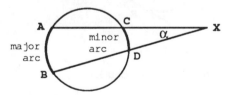

When the point X lies outside the circle, the above formulas have **limiting cases.** By a limiting case, we mean the formula we get when we let one or both of the lines through X go to the very edge of the circle. (Neither X nor the circle move, the line just rotates.) The line XE which meets the circle in just one point is called a **tangent** to the circle.

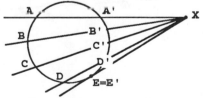

(We will figure out formulas for the limiting cases in the exercises.)

1. **Explain why**
   $|XA| \cdot |XA'| = |XE|^2$.

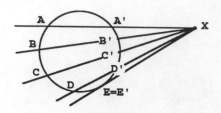

2. **Explain why**
   $\alpha = \frac{1}{2}|\overset{\frown}{AE}| - \frac{1}{2}|\overset{\frown}{A'E}|$.

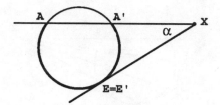

3. **Explain why**
   $|XA|^2 = |XE|^2$.

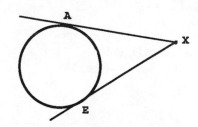

4. **Explain why**
   $\alpha = \frac{1}{2}|\text{major}\overset{\frown}{AE}|$
   $\qquad - \frac{1}{2}|\text{minor}\overset{\frown}{AE}|$.

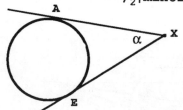

5.

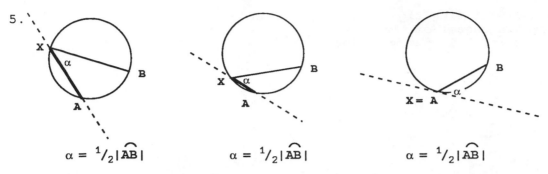

$$\alpha = \frac{1}{2}|\overset{\frown}{AB}| \qquad \alpha = \frac{1}{2}|\overset{\frown}{AB}| \qquad \alpha = \frac{1}{2}|\overset{\frown}{AB}|$$

In the above "moving picture," A and B don't move, but X moves down along the circle until it gets to A. The measure of the angle α doesn't change because, by I25, it is always equal to $\frac{1}{2}|\overset{\frown}{AB}|$

which doesn't change. The last picture tells us how to calculate the angle between a tangent line and a chord. State this formula as precisely as you can.

6. If the line L is tangent at A to the circle with center O, explain why L is perpendicular to the line OA.

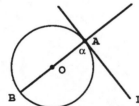

7. Explain why

ΔABC

is isosceles.

8. In the following figure,
   O is the center of the
   circle and M is the midpoint
   of OP.

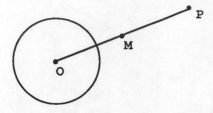

Draw the circle with center M
and radius |MO| (which is the
same as |MP|).

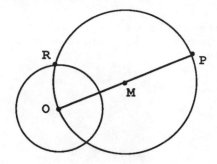

Show that ∠ORP is 90°. Use
exercise #6 to show that
the line through P and R is
tangent to the circle with
center O.

Degrees and radians

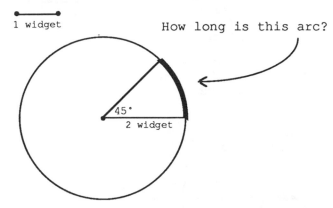

1 widget

How long is this arc?

The circumference
of the circle
is $2\pi r = 2\pi \cdot 2 = 4\pi$ widgets.

Since the arc is cut by
a central angle of 45˚,
which is 1/8 if the 360˚
in the whole circle, the
length of the arc must be
1/8 of the circumference,
that is, it must be
$(1/8)\cdot 4\pi$ widgets, or simply
$\pi/2$ widgets.

```
length of arc = ((central angle)/360˚)·2πr
              = ((central angle)/360˚)·2π "radiuses"
```

$$\frac{\text{length of arc}}{\text{radius of circle}} \quad = \quad \left(\frac{\text{central angle}}{360˚}\right) 2\pi$$

So the ratio of the length of the arc to the radius of the circle
depends only on the size of the central angle. This ratio is
called the measure of the central angle in **radians**.

```
measure in radians = ((measure in degrees)/360)·2π

measure in degrees = ((measure in radians)/2π)·360
```

Maybe the easiest way to remember both of these formulas is simply
to remember:

$$\frac{\textbf{measure in radians}}{2\pi} \quad = \quad \frac{\textbf{measure in degrees}}{360}$$

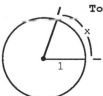

To construct an angle of x radians:

Cut a piece of string x units long.
Draw a circle of radius 1 unit around (0,0) in the (x,y)-plane.
Starting from the point (1,0), wrap the string around the circle.
Mark the point on the circle A where the string ends.
The angle of x radians is the angle made by the positive
x-axis and the ray from (0,0) through A.

1. Give the radian measure
    for the following angles:

    180°   _____

    90°   _____

    45°   _____

    30°   _____

    60°   _____

    120°   _____

    270°   _____

    15°   _____

    1°   _____

2. Give the degree measure
    for the following angles:

    π radians

    $\pi/2$ radians   _____

    $\pi/3$ radians   _____

    1 radian   _____

    $(3/2)\pi$ radians   _____

    $\pi/6$ radians   _____

    $(2/3)\pi$ radians   _____

    0 radians   _____

3. When the radian measure
    of an angle gets up to
    2π, the string we wrap
    around the unit circle
    ends back at the point
    (1,0). So this angle is
    the same as the angle
    of 0 radians. On the
    unit circle at the right,
    show the angle of $(5/2)\pi$
    radians. Find a radian
    measure for this angle
    which is less than 2π
    radians.

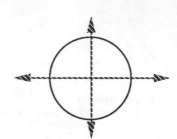

4. Find a radian measure
    which is less than 2
    for each angle:

    7π radians
    $(8/3)\pi$ radians
    1001π radians
    2π radians

5. We will also let
    radian measures be
    negative. For example,
    to construct an angle
    of $-(\pi/3)$ radians, cut
    a piece of string $(\pi/3)$
    units long. Starting from
    (1,0), instead of wrapping
    the string in the upward
    direction, we wrap in the
    <u>downward</u> direction.
    Draw in the angles of
    given measures:

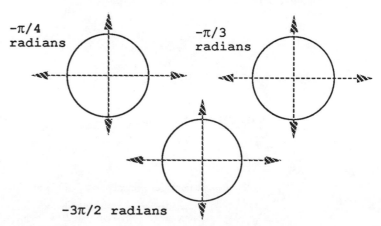

$-\pi/4$ radians $-\pi/3$ radians

$-3\pi/2$ radians

Trigonometry

Suppose I have an angle:

All trigonometry is based on this simple fact:

All right triangles which have the angle α as one of their (other) angles are similar.

To see this, let's compare two right triangles that have ∠α as one of their angles.

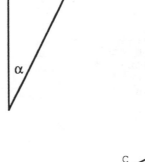

Rotate the second triangle
around and flip it over to
get its angle α to lie on top
of angle α of the first triangle,
and so that the right angles
of the two triangles
lie on the same
side of the
angle α:

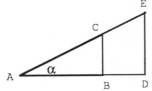

Now BC is parallel to DE by the Converse of the Z-principle, so by exercise #1 of I20e, ΔABC ~ ΔADE. So the two triangles I started with are similar, since I didn't bend, stretch or break the second triangle when I was moving it into position.

So, by Condition Q of I20: |AB|/|AD| = |AC|/|AE|
So, by algebra: |AB|/|AC| = |AD|/|AE|

This means that the ratio of the length of the
adjacent side to the length of the hypoteneuse
is the same for all right triangles with angle α:

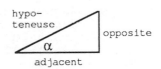

So this ratio depends only on α. It is called **cosine α**.
We reason in the same way for the other ratios:

$$\text{sine } \alpha = \frac{\text{opposite}}{\text{hypoteneuse}} \qquad \text{tangent } \alpha = \frac{\text{opposite}}{\text{adjacent}} \qquad \text{secant } \alpha = \frac{\text{hypoteneuse}}{\text{adjacent}}$$

$$\text{cosine } \alpha = \frac{\text{adjacent}}{\text{hypoteneuse}} \qquad \text{cotangent } \alpha = \frac{\text{adjacent}}{\text{opposite}} \qquad \text{cosecant } \alpha = \frac{\text{hypoteneuse}}{\text{opposite}}$$

1.

sine α = _____
cosine α = _____
tangent α = _____
cotangent α = _____
secant α = _____
cosecant α = _____

2. Find sine 45°:
   (Hint: Any right
   triangle with
   an angle of
   45° is isosceles.
   Then use the Pythagorean
   theorem to find x.)

"rad." means
"radians"

3. Find cosine 45°   _____
        tangent 45°   _____
        cotangent 45°   _____
        secant 45°   _____
        cosecant 45°   _____

4. Find cosine (π/4 rad.)   _____
        tangent (π/4 rad.)   _____
        cotangent (π/4 rad.)   _____
        secant (π/4 rad.)   _____
        cosecant (π/4 rad.)   _____

5. Find sine 30°.
   (Hint: 2α = 60°
   since each
   angle of an
   equlateral
   triangle has
   measure 60°)

6. Find cosine 30°
   (Hint; Use the
   Pythagorean
   theorem to
   find x.)

7. Find cosine (π/6 rad.).

8. Find tangent 30°   _____
         cotangent 30°   _____
         secant 30°   _____
         cosecant 30°   _____

   Find tangent (π/6 rad.)   _____
        cotangent (π/6 rad)   _____
        secant (π/6 rad.)   _____
        cosecant (π/6 rad.)   _____

9. A regular n-gon is a polygon with n sides, such that the lengths of all its sides are equal and the measures of all its interior angles are equal. (See C8-C11.) A circle is <u>inscribed</u> in a polygon if it lies inside the polygon and is tangent (I25) to each side of the polygon. A regular n-gon admits an inscribed circle and a radius of that circle, drawn to one of the points of tangency is called an <u>apothem</u> of the regular n-gon. A circle is <u>circumscribed</u> about a polygon if it passes through all the vertices of the polygon. A regular n-gon admits a circumscribed circle and a radius of the circle , drawn to a vertex to the polygon is called a <u>radius</u> of the polygon.
In this exercise, we study regular n-gons whose sides are all of length s, with apothem a and radius r. Explain why, for regular hexagons (n=6):

$a = r \cdot \sin 60°$ $a = (s/2) \tan 60°$.

For regular pentagons

$a = r \cdot \sin \underline{\quad}°$ $a = (s/2) \tan \underline{\quad}°$

For regular n-gons:

$a = r \sin \underline{\quad}°$ $a = (s/2) \tan \underline{\quad}°$

10. Fill in the following table for regular n-gons:

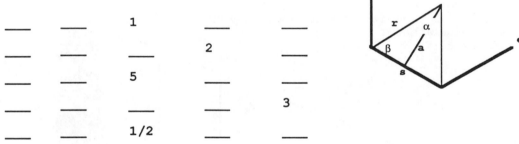

n=	α=	β=	s=	r=	a=
6	__	__	1	__	__
5	__	__	__	2	__
3	__	__	5	__	__
4	__	__	__	__	3
7	__	__	1/2	__	__

(Use your calculator.)

Tangent α = (rise)/(run)

All of the triangles in the picture
below are similar. So, The ratio
of a rise to its corresponding run
depends <u>only</u> on the size of
the angle α. This ratio
is called the tangent
of the angle, and is
written "tangent α".

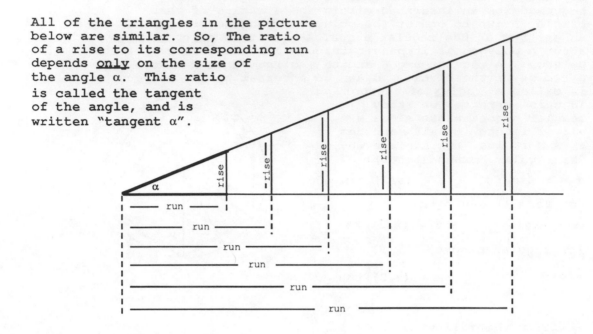

We can use this fact to find the height
of a tree without leaving
the ground:

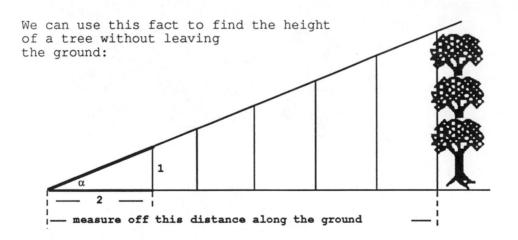

— **measure off this distance along the ground**

$$\frac{1}{2} = \frac{\text{height of tree}}{\text{distance along ground}} \qquad \text{height of tree} = \frac{1}{2} \text{ distance along ground}$$

height of tree = (tangentα)(distance along ground)

1. **In the formula at the bottom of I28, your calculator will give you the value of tangent α once you know the angle α (either in degrees or radians). Practice using this function of your calculator:**

> Warning: Make sure that your calculator is in "degree" mode if you are entering degrees, and in "radian" mode if you are entering radians.

tangent 30° = _____

tangent 51° = _____

tangent 45° = _____

tangent (–15°) = _____

tangent 89° = _____

tangent (π/4 rad.) = _____

tangent (1 rad.) = _____

tangent (0.5 rad.) = _____

tangent (–π/4 rad.) = _____

tangent (49π/100 rad.) = _____

2. **Your calculator will also give you the measure of an angle (in either degrees or radians), if you know the value of the tangent of the angle. Just use the "INV TAN" function. Practice this:**

> Warning: Make sure that your calculator is in "degree" mode if you want to get your answer in degrees, and in "radian" mode if you want your answer in radians.

tangent α = 1   α = _____ °   α = _____ rad.

tangent α = 1.732   α = _____ °   α = _____ rad.

tangent α = 0.333   α = _____ °   α = _____ rad.

tangent α = –1   α = _____ °   α = _____ rad.

tangent α = 1.414   α = _____ °   α = _____ rad.

3. **How tall is this tree? (Hint: Use your calculator.)**

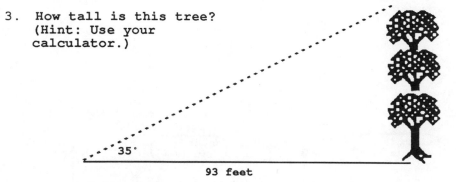

35°

93 feet

4. **What is the radian measure of the angle α ?**

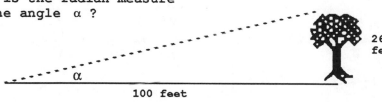

26 feet

α

100 feet

Everything you always wanted to know about trigonometry but were afraid to ask

This picture, the Pythagorean theorem, and a little brainpower will get you almost all the trigonometry you'll ever need!

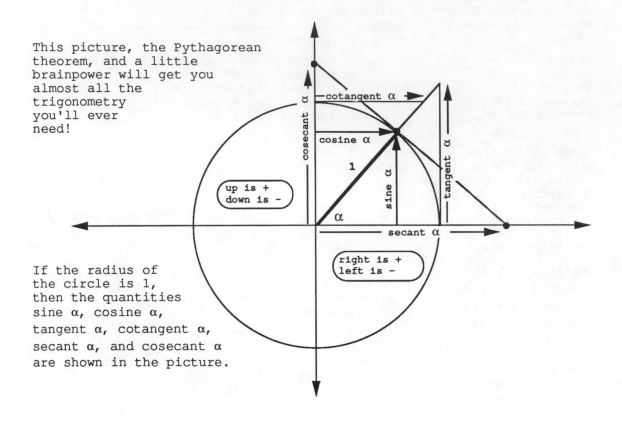

up is +
down is -

right is +
left is -

If the radius of the circle is 1, then the quantities sine α, cosine α, tangent α, cotangent α, secant α, and cosecant α are shown in the picture.

The angle can pass 90° = π/2 rad.,   or 180° = π rad.,   or even 270° = 3π/2 rad.

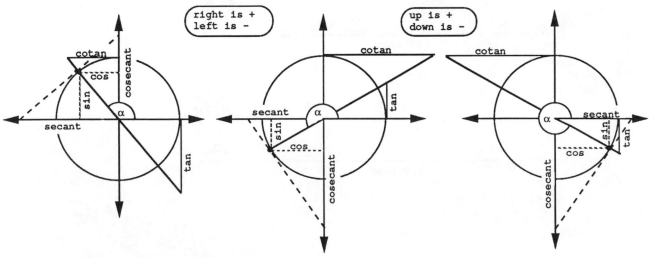

1. Explain why

   $(\text{sine } \alpha)^2 + (\text{cosine } \alpha)^2$

   is always equal to one,
   no matter what the
   angle α is. (Hint:
   Use the Pythagorean
   theorem.)

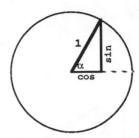

2. If cosine $\alpha = 0.654$,
   what is sine α ?

3. Explain why

   $$\text{tangent } \alpha = \frac{\text{sine } \alpha}{\text{cosine } \alpha}$$

   no matter what the
   angle α is.

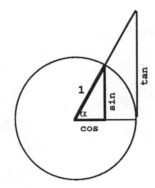

4. If cosine $\alpha = 0.654$,
   what is tangent α?
   (Hint: Use exercise #2.)

5. Use facts that
   you know about
   congruent
   triangles and
   the picture at
   the right to
   show that

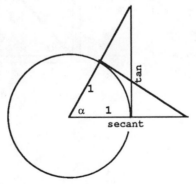

   $1 + (\text{tangent } \alpha)^2$
   $\quad = (\text{secant } \alpha)^2$

   no matter what the angle α is.

6. If tangent α = 1.156
   what is secant α ?

7. Use what you know
   about similar triangles
   in the picture
   at the right
   to explain
   why

   secant α =

   1/(cosine α)

   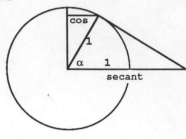

   no matter
   what the
   angle α is.

8. If cosine α = 0.654,
   what is secant α ?

9. Look at the picture
   in I29, and write down
   three other relations
   (like exercises #1,#3,
   #5 and #7) that hold
   no matter what the
   angle α is.

10. Explain why
    cos α = sin(90°−α)
    cotan α = tan(90°− α)
    cosec α = sec(90°− α)
    using the picture in I29.

    ┌─────────────────────────────────┐
    │ cosine of angle =   │
    │   sine of complementary angle   │
    │ cotangent of angle =   │
    │   tangent of complementary angle│
    │ cosecant of angle =   │
    │   secant of complementary angle │
    └─────────────────────────────────┘

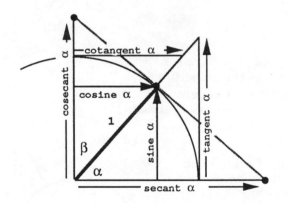

11. Using the picture in I29,
    or exercise #1, explain
    why

    −1 ≤ sine α ≤ 1
    −1 ≤ cosine α ≤ 1

    no matter what the angle
    α is.

The law of sines and the law of cosines

Any
 triangle:

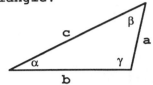

Law of sines:

$$\frac{\text{sine } \alpha}{a} = \frac{\text{sine } \beta}{b} = \frac{\text{sine } \gamma}{c}$$

To see why the law of sines is true, draw in a segment
through C which is perpendicular to AB at some point D:

$\sin \alpha = \frac{\text{opp.}}{\text{hyp.}} = \frac{|CD|}{|AC|}$ $\sin \beta = \frac{\text{opp.}}{\text{hyp.}} = \frac{|CD|}{|BC|}$

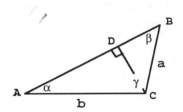

So by algebra: $|AC| \sin \alpha = |CD| = |BC| \sin \beta$

$$\frac{\sin \alpha}{|BC|} = \frac{\sin \beta}{|AC|} \quad \text{that is,} \quad \frac{\sin \alpha}{a} = \frac{\sin \beta}{b}$$

Any
 triangle:

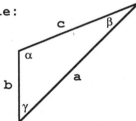

Law of cosines:

$$c^2 = a^2 + b^2 - 2ab \cdot \cos \gamma$$

To see why the Law of Cosines is true, construct a
segment through A which is perpendicular to BC at
some point D:

The Pythagorean theorem says:

$$|CD|^2 + |AD|^2 = b^2 \qquad \text{and} \qquad |BD|^2 + |AD|^2 = c^2$$

Now do some algebra with these two equations:

$c^2 = (b^2 - |CD|^2) + |BD|^2 \ = b^2 + |BD|^2 + 2|BD||CD| + |CD|^2 - 2|BD||CD| - 2|CD|^2$

$= b^2 + (|BD|+|CD|)^2 - 2|CD|(|BD|+|CD|) = b^2 + a^2 - 2a|CD|$

But $\cos \gamma = (\text{adj.})/(\text{hyp.}) = |CD|/b$,   so $|CD| = b \cdot \cos \gamma$.   Now just substitute
for $|CD|$ in the equation

$$c^2 = b^2 + a^2 - 2a|CD|$$

to get the formula we want.

1. In these exercises, we will
   use the laws of sine and
   cosine to "figure triangles."
   For example, we know by SSS
   that once the lengths of the
   three sides of a triangle are
   given, the triangle is
   completely determined. So we
   ought to be able to figure
   out the angles. Use the Law
   of Cosines and the INV COS
   function on your calculator
   to figure the three angles
   in the triangle below:

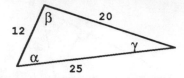

2. SAS: Figure (the remaining
   sides and angles of) the
   triangle below:

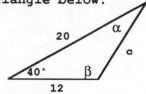

3. You used the Law of Cosines in
   exercise #2 to calculate c =
   13.3. Then you might have used
   Law of Cosines or Law of Sines for
   the rest. Use the Law of Sines

$$\frac{\sin \beta}{20} = \frac{\sin 40°}{13.3}$$

   to calculate sin β. Now use INV
   SIN to calculate β. On the other
   hand, you might have used the
   Law of Cosines to calculate cosβ :
   $20^2 = 12^2 + 13.3^2 - 2 \cdot 12 \cdot 13.3 \cdot \cos \beta$
   Now use INV COS to calculate β.
   Explain why you don't get the
   same answer. Which answer is
   correct?

4. **ASA: Figure the triangle below:**

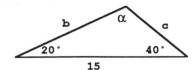

(Hint: Use Law of Sines.)

5. **AAS: Figure the triangle below:**

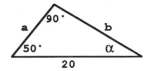

6. **Why ASS doesn't quite work:**

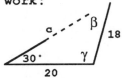

Use the Law of Cosines to see that there are two possible values for c.

7. In the Law of Cosines in I30, suppose $\gamma = 90°$, that is, $\gamma = \pi/2$ radians. What does the Law of Cosines become in that case?

8. The following is true of
    the diagram below:

    • ∠RPO, ∠QNO, and ∠PMO
        are right angles.
    • |QN| = sin α
    • |OQ| = 1
    • |OP| = cos β
    • |ON| = cos α
    • |RP| = sin β

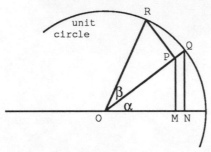

    Explain why |PM| = sin α ·cos β.
    (Note that α and β could be
    any angles.)

9. Now explain why
    ∠URP = ∠α. Once this
    is shown, the following
    conclusion can be drawn:

    Because ΔRUP is a right triangle:

    $$\cos \angle URP = \frac{|UR|}{|RP|}$$

    $$\cos \alpha = \frac{|UR|}{\sin \beta}$$

    |UR| = cos α·sin β

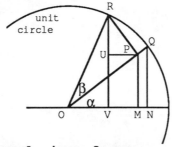

10. Use the conclusions from
    ex.#8,9 to explain why, for
    any angles α and β,

    sin (α+β)

        = sin α cos β + cos α sin β

Figuring areas

In I30, we saw that if we know SSS or SAS or ASA or AAS for a triangle, we can figure out the rest of the sides and angles of the triangle. We ought to be able to figure out the <u>area</u> of the triangle too.

Remember area? Choose a unit: 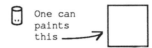

It takes one can of paint to paint a square one widget on a side:

One can paints this ⟶ ▢

SAS: How many cans of paint does it take to paint this triangle:

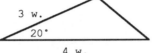

First draw in the height h:

$$\sin 20° = {}^h/_3$$
$$0.342 = {}^h/_3 \quad h = 1.026 \text{ w.}$$

So, area = (1/2)bh = (1/2)x4x1.026 = 2.052 square widgets, that is, it will take 2.052 cans of paint to paint the triangle.

ASA or AAS: How many cans of paint does it take to paint the triangle:

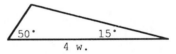

Use I30 to find a second side, then use SAS method.

SSS: How many cans of paint does it take to paint this triangle:

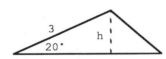

We saw above that: area = (1/2)x2x4sin α

But: $(\sin α)^2 + (\cos β)^2 = 1.$

So   area = $(1/2)x2x4 \sqrt{1-(\cos α)^2}$

Now by the Law of Cosines:

$$3^2 = 2^2 + 4^2 - 2x2x4\cos α$$

So:

$$\cos α = (3^2\ 2^2 - 4^2)/(-2x2x4)$$

By substitution, then:

area =

$$(1/2)x2x4 \sqrt{1-[(3^2-2^2-4^2)/(-2x2x4)]^2}$$

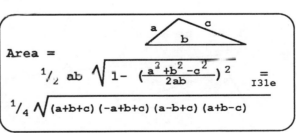

$$\text{Area} =$$
$$\tfrac{1}{2}\ ab\ \sqrt{1- (\tfrac{a^2+b^2-c^2}{2ab})^2}\ \overset{=}{}_{\text{I31e}}$$
$$\tfrac{1}{4}\ \sqrt{(a+b+c)(-a+b+c)(a-b+c)(a+b-c)}$$

There's some algebra involved in the last equality in this box. We'll do it in the exercises.

1. Find the area of the triangle with sides 11, 20 and 30 widgets.

2. Find the area of the triangle with sides 10, 20 and 30 widgets. Explain your answer.

3. Find the area of the triangle with sides 9, 20 and 30 widgets. Explain your answer.

4. Finish the algebra in the SSS formula for the area of a triangle in I31:

$$\text{Area} = \frac{1}{2}ab \sqrt{1 - \left(\frac{a^2+b^2-c^2}{2ab}\right)^2}$$

= ----------------------------

$$= \frac{1}{4}\sqrt{(2ab)^2\left[1 - \left(\frac{a^2+b^2-c^2}{2ab}\right)^2\right]}$$

= ----------------------------

$$= \frac{1}{4}\sqrt{(2ab+(a^2+b^2-c^2))(2ab-(a^2+b^2-c^2))}$$

= ----------------------------

= ----------------------------

$$= \frac{1}{4}\sqrt{(a+b+c)(-a+b+c)(a-b+c)(a+b-c)}$$

5. Find the area of the triangle below:

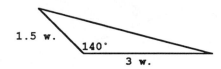

6. Find the area of the triangle below:

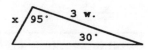

7. If the diagonals of a
   quadrilateral meet inside
   the quadrilateral

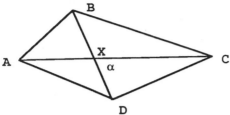

   show that

   area ABCD =
          $(^1/_2)$ |AC|·|BD|·sin α.

   (Hint: Write
   |AC| = |AX| + |XC|
   |BD| = |BX| + |XD|
   and use that
   sin α = sin (180° − α).)

8. In the diagram below, show that
   |AB|/|CB| = sin α /sin β:

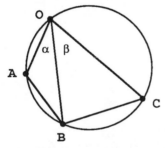

   (Hint: Use I24 to show that
   sin ∠OAB = sin ∠OCB.)

9. In the diagram below, show that
   |A'B'|/|C'B'| =
   (sin α/sin β)/(sin γ/sin δ):

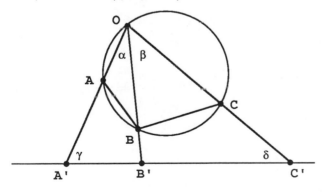

10. In the diagram below, show that
    (|AB|/|CB|)/(|AD|/|CD|) =
    (|A'B'|/|C'B'|)/(|A'D'|/|C'D'|).

    (|AB|/|CB|)/(|AD|/|CD|) is called
    the <u>cross-ratio</u> of (A,B,C,D).

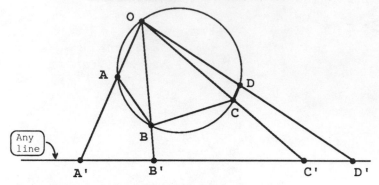

    (Hint: Use ex.#8 and #9 twice
    each, once with D replacing B,
    once with D' replacing B'.)

11. Use the cross-ratio to prove
    Ptolemy's theorem:

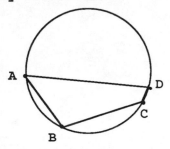

    If A, B, C, and D lie on a
    circle in the order shown,
    then

    |AC|·|BD| =
        |AD|·|BC| + |AB|·|CD|.

    (Hint: Divide both sides of
    the equation by |AD|·|BC|
    and write the two quotients
    which appear as
    cross-ratios. Use exercise
    #10 to replace these
    quotients by cross-ratios on
    a line, e.g. the line
    through A and D:

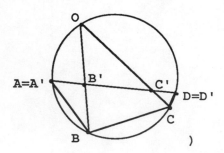

)

The second magnification principle

Back in I16, we studied the First Magnification Principle:

If an n-dimensional figure is magnified in all directions by a factor of r, its (length, area, volume) is multiplied by a factor of r ^{n}.

Now let's see what happens if we stretch different directions by <u>different</u> factors. For example, in the (x,y)-plane, suppose we magnify in the x-direction by a factor of 3 and in the y-direction by a factor of 2. Then a funny thing happens to lengths:

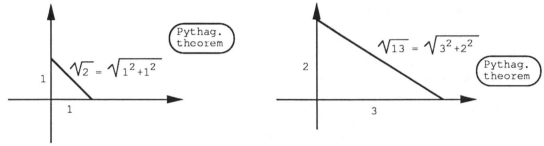

Horizontal lengths are tripled and vertical lengths are doubled but slant lengths change by weird amounts! However, the area of the triangle goes from (1/2)x1x1 to (1/2)x3x2, that is, the area gets multiplied by 3x2.

This works for the area of any blob (that is, shape) in the plane:

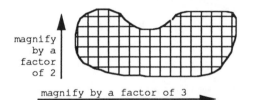

magnify by a factor of 2

magnify by a factor of 3

We can (approximately) divide the blob up into little squares.

Each little square will be magnifified by a factor of 3 in the horizontal direction and by a factor of 2 in the vertical direction—so each little square will have its area multiplied by a factor of 6:

□ ⟶ ▭

The whole magnified figure will be made up of the same number of pieces (rectangles instead of squares) as before, but each piece will have 3x2=6 times the area it had before. So the area of the magnified figure is obtained by multiplying the area of the old figure by 6.

The same reasoning works for volumes in space:
Divide the solid figure up (approximately) into little cubes. If we stretch space by a factor of 3 in the x-direction, by a factor of 2 in the y-direction and by a factor of 5 in the z-direction, the volume of each little cube gets multiplied by a factor of 3x2x5=30. So the volume of the whole solid figure is 30 times what it was before.

If an n-dimensional figure <u>in n-dimensional space</u> is magnified in perpendicular directions by factors of r_1, r_2, ..., r_n, its (n-dimensional) volume is multiplied by a factor of
$$r_1 \cdot r_2 \cdot \ldots \cdot r_n.$$

1. A circle of radius 1
    is stretched horizon-
    tally by a factor of
    3 and vertically by a
    factor of 5. What is
    the area inside the
    magnified figure?
    (The stretched circle
    is called an <u>ellipse</u>.)

2. A circle of radius 1
    around (0,0) in the
    (x,y)-plane is stretched
    by a factor of 3 in the
    x-direction and shrunk
    by a factor of (1/2) in
    the y-direction. Graph
    the resulting ellipse at
    the right, and find the
    area it encloses.

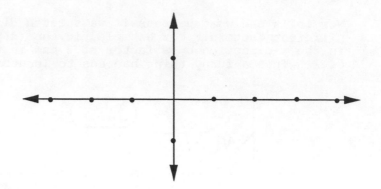

3. The pyramid at the right
    is stretched in the x-
    direction by a factor
    of 4 and in the y-
    direction by a factor
    of 2, while its height
    remains unchanged.
    By what factor is its
    volume multiplied?

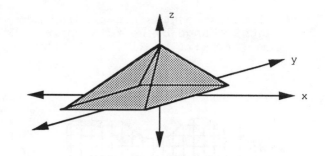

4. The area of this triangle
    is (1/2):

    Use the Principle of
    Parallel Slices (I13)
    and the Second Magni-
    fication Principle to
    give a new way to see
    that the formula for
    the area of a triangle
    is   (1/2)·b·h.

5. **After learning the Second Magmification Principle for squares, we said that it also worked for any blob in the plane:**

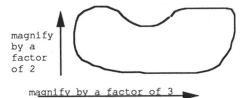

magnify by a factor of 2

magnify by a factor of 3

We did this by dividing the blob up approximately into little squares:

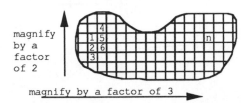

magnify by a factor of 2

magnify by a factor of 3

Complete the argument at the right.

The area of the blob is approximately the sum of the areas of the little squares which fit entirely inside the blob, that is,
(area 1)+(area 2)+ ... +(area n).
Now do the 3x2 magnification.
The magnification multiplies the area of each little square by a factor of 6 (=2x3). Altogether we get
6(area 1) + 6(area 2) + ... + 6(area n),
which, by the _____ law, is equal to
6[(area 1)+(area 2)+ ... +(area n)].
But this last quantity is
6(approximate area of blob).
So the magnification multiplies the approximate area of the blob by a factor of ___.

6. **The reasoning in exercise #5 is only approximate. Get a better approximation by dividing each of the little squares into even smaller squares:**

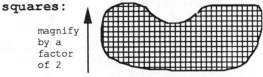

magnify by a factor of 2

magnify by a factor of 3

Explain why the sum of the areas of all the tiny squares lying entirely inside the blob is a better approximation to the area of the blob than the sum of the areas of the squares in exercise #5.

7. **Repeat the argument in exercise #5 with the better approximation to the blob we used in exercise #6.**

Volume of a pyramid

Any figure in
the plane

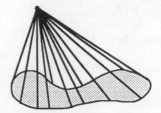

Build a pyramid over the
figure to some point above it

The question we want to answer is: What is the volume of the three-
dimensional pyramid? That is, if it takes 1 bucket of glop to fill
a cube 1 widget by 1 widget by 1 widget, how many buckets of glop
will it take to fill the (inside of) the pyramid?

Start with
a 1x1x1 cube:

Pick a vertex (corner)
of the cube. How many
faces (square sides)
of the cube don't
contain the corner
you chose?

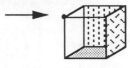

Make a pyramid
from the chosen
vertex to each
of the 3 faces
which don't contain
the vertex.

Those three pyramids are
all the same and together
they fill up the cube.
So, if it takes 1 bucket
of glop to fill the cube
it will take 1/3 bucket
of glop to fill each pyramid!

Volume = (1/3) cubic widgets
              (cu.w.)

So, by the Second Magnification
Principle, the volume of the
pyramid at the right is:

$$r_1 \cdot r_2 \cdot r_3 \cdot (1/3) \text{ cu.w.}$$

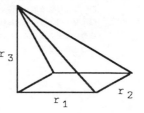

All these pyramids below have the same volume as the one we just did.
That's because of the Principle of Parallel Slices (I13), this time in 3 dimensions.
We just slice our figure in "slabs" parallel to the base and slide it over without
changing the volume of any slab:

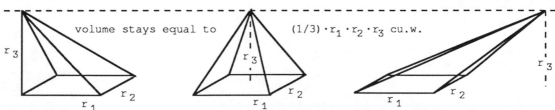

volume stays equal to $(1/3) \cdot r_1 \cdot r_2 \cdot r_3$ cu.w.

Volume of pyramid = (1/3)(area of base)(vertical height)

So far we have only reasoned out this formula in the case in which
the base of the pyramid is a rectangle. In the exercises, we will see
that the formula is true in general, that is, when the base of the pyramid
is <u>any</u> blob in the plane.

1. A pyramid is built
   over a rectangular
   base 3 widgets by
   4 widgets. Its
   vertical height is
   15 widgets. What is
   its volume?

2.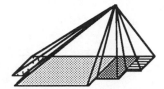

   A pyramid is built
   over a shape made up
   of four rectangles
   as shown above. The
   rectangles have areas
   1, 8, 2.5, and 0.75
   sq.w. respectively.
   The pyramid has vertical
   height 1.5 widgets. What
   is its volume?

3. In exercise #2, you can
   figure out the volume of
   each of the four pieces of
   the pyramid separately, then
   add, or you can add the areas
   of the four rectangles first,
   then multiply by (1/3) and by
   the vertical height. What
   law of arithmetic are you
   using to conclude that you
   get the same answer either way?

4. Explain why the volume of the
   pyramid below is
   (1/3)(area of base)(vert.height).

   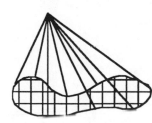

   (Hint: Cut the base
   up approximately
   into rectangles
   and use the
   distributive law.)

5. A one-widget vertical
    segment moved along a
    one widget horizontal
    segment gives a square:

    Choose a vertex
    (corner) of the
    square:

    The chosen vertex is an
    endpoint of the segment
    above A and of the
    segment obtained by moving
    the chosen vertex along above
    the arrow from A to B. So
    the chosen point lies in two
    edges of the square.

    Think of a 1x1x1 cube as
    being scraped out by a
    square moving along for
    one unit in a direction
    perpendicular to the square:
     Choose a vertex
    and repeat the
    above reasoning
    to figure out that
    the vertex lies on
    three faces of the
    cube.

6. Choose a vertex of the 1x1
    square, and make a
    "2-dimensional pyramid"
    (triangle) by filling in
    all possible segments between
    the vertex and points on
    an edge of the square which
     does **not** contain
    the vertex.
    Since there are two
    edges of the square
    not containing the
    chosen vertex and
    the 1x1 square has
    area 1, the area of each
    triangle must be ____.

7. Choose a vertex of the 1x1x1
   cube, and make a
   "3-dimensional pyramid"
   by filling in all possible
   segments between
   the vertex and points on
   a face of the cube which
   does <u>not</u> contain
   the vertex.
   Since there are three
   faces of the cube
   not containing the
   chosen vertex and
   the 1x1x1 cube has
   volume 1, the volume of each
   pyramid must be ____.

8. A four-dimensional
   cube can be thought
   of as a 1x1x1 three-
   dimensional cube
   moved along a segment
   of length one in a
   direction "perpendicular"
   to the 3-dimensional cube.
   If a square has 4
   faces, each a segment,
   a 3-dimensional cube
   has 6 faces, each a
   square, how many faces
   does a 4-dimensional
   cube have? (Each is
   a 3-dimensional cube.)

9. Pick a vertex (corner) of
   the 4-dimensional cube.
   How many 3-dimensional
   faces of the four dimensional cube do not touch
   the chosen vertex? Sweep
   out a 4-dimensional pyramid
   of height 1 and base 1x1x1
   by connecting the chosen
   vertex to all points of an
   opposite face. If the volume
   of the 4-dimensional cube is 1,
   what is the (4-dimensional)
   volume of this pyramid?

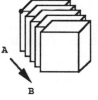

10. Guess the formula for the
    volume of a 4-dimensional
    pyramid.

Of cones and collars

Suppose we take a circle with center O and lay it down flat in
the plane:

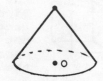

Next make a pyramid whose vertex is directly above O:

The outside surface of the pyramid, not including
the base circle, is called a **(right circular) cone.**

How many cans of paint does it take to paint the cone?
(Remember, 1 can of paint paints a square 1 widget by 1 widget.)

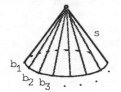

To figure this out, slice the cone up into
pieces that are (almost) thin triangles:

The base of the first triangle is (approximately) b_1 ,

the base of the second triangle is (approximately) b_2 ,

the base of the third triangle is (approximately) b_3 ,

                      etc.

The altitude, or height, of each triangle is (approximately)
the **slant height** s of the cone.

So the area is approximately $(1/2)b_1 \cdot s + (1/2)b_2 \cdot s + (1/2)b_3 \cdot s + \ldots =$

$$(1/2)(b_1 + b_2 + b_3 + \ldots) \cdot s$$

Surface area of right circular cone =
            (1/2)(circumference of base)(slant height)

Now cut the top off a (right circular) cone:
We'll call the "band" that is left a **collar.**

Suppose now we want to just paint the collar.
How much paint will it take?

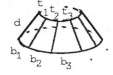

We cut the collar into pieces, like we did before.
Each piece is (approximately) a trapezoid. By exercise #2
of I13e, the area of the first trapezoid is $(1/2)(b_1 + t_1)d$

The area of the second trapezoid is $(1/2)(b_2 + t_2)d$

The area of the third trapezoid is $(1/2)(b_3 + t_3)d$

          etc. $\ldots$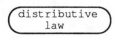

Total area $(1/2)(c_b + c_t)d$

    where c_b = circumference of bottom circle   and c_t = circumference of top circle.

Surface area of right circular collar =
(1/2)·[(circum. bottom circle)+(circum. top circle)]·(slant height)

1. A right circular cone has
   as base a circle of radius
   2 widgets and slant height
   of 3 widgets. Find its
   surface area. (We do not
   include in this the area
   of the base.)

2. A right circular cone has
   a base of radius 2 and a
   <u>vertical</u> height of 3. What
   is its surface area? (Hint:
   Use the Pythagorean theorem
   to find the slant height.)

3. Find the surface area of
   a collar whose top circle
   has circumference 3 w. and
   whose bottom circle has
   circumference 4 w. and
   whose slant height is 1 w.

4. Find the volume of the
   (inside of) the cone in
   exercise #1. (Hint: A
   cone is a special type
   of pyramid.)

5. Suppose the right circular
   cone in exercise #1 is sliced
   by a plane parallel to the
   base halfway between the
   base and the vertex. Find
   the surface area of the
   part of the cone below
   the plane (not including
   the area of the base).

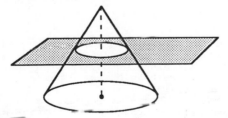

Two planes
are parallel
if they
never meet.

Sphereworld

We are now ready to try some 2-dimensional geometry in "another universe" instead of in the plane. Our new 2-dimensional universe will be the surface of a **sphere**:

A sphere of radius r is the set of all points in 3-dimensional space whose distance from a given point O is exactly r.

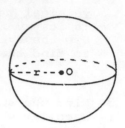

The number **r** is called the **radius** of the sphere.
The point **O** is called the **center** of the sphere.

(You should compare there definitions with the definitions at the bottom of I2.)

For the time being, we will pretend that we can only move around on the surface of the sphere and that we are very, very small compared to the radius of the sphere. In fact, that is usually the state of affairs in our daily lives— we call the sphere the (surface of the) earth.

Line   1. A line extends infinitely in two directions.
2. Given any two distinct points in sphereworld, there is a line containing both of them.

(Compare with I1)

3. Given any two points on a line, the quickest way to travel from one to the other is to stay on the line.
4. Take any two points out of a line. What's left has two separate pieces.

Mathematicians have figured out that, in sphereworld, the lines are constructed as follows:
> Take a plane in 3-dimensional space which passes through the center of the sphere. The line is the set of points along which the plane cuts the sphere.

Lines in sphereworld are also called **great circles**.

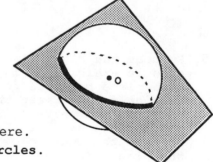

Some comments on the four properties of lines in sphereworld which we listed above:

1. Lines "go on forever" but, if we keep going in one direction along the line, eventually we get back where we started from. If we keep going in the same direction after that, we just retrace our steps again and again.
2. Given any two points P and Q in sphereworld, we saw in I2 that we can find a plane that contains the three points P, Q, and O. Intersecting that plane with the sphere, we get the line in sphereworld passing through P and Q.
3. The fact that the shortest distance between P and Q lies along the great circle through P and Q is used by intercontinental airlines all the time. Look on a globe and trace out the shortest route from San Francisco to London.
4. Since lines in sphereworld "circle back on themselves," we cannot disconnect a line by just taking out one point. But lines are one-dimensional paths, so if we take out two (or more) points, we break the line into pieces. What we want to say with this property is that lines aren't "thick." (See exercise #3 of I1e.)

1. Given two points, P and Q, in sphereworld, is there always exactly one line (great circle) through P and Q? If not, when is there more than one line through P and Q? (Hint: How many planes contain P, Q, and O, the center of the sphere?)

2. We will say that two lines (great circles) meet perpendicularly if the planes through O which cut out the circles meet perpendicularly. Decide what it should mean for two planes in space to meet perpendicularly:

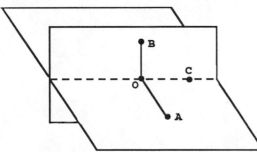 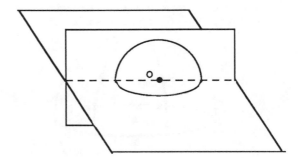

Construct

OA ⊥ OC in the first plane,

OB ⊥ OC in the second plane.

Then the planes are perpendicular if ∠AOB = _____°.

3. Using a globe, give an example of perpendicular lines (great circles) in sphereworld.

4. Now suppose we have a
   sphere with radius 1
   unit. (If our units are
   widgets, the sphere would
   be about the size of a
   golf ball; if the units are
   meters, you could barely get
   the sphere into the back of
   a small truck; if the units
   are astronomical units, all
   of the solar system out to
   the earth would fit inside
   the sphere.) How many units
   long is a line in sphereworld?

5. Suppose you stand at a point
   P in the northern hemisphere
   of the earth, walk a units
   directly south to the equator,
   turn 90° to the left,
   walk b units along the
   equator, turn 90° to the left,
   walk a units toward the
   north pole, and finally
   turn 90° to the left,
   and walk b units along a
   line (great circle). Do
   you get back to the original
   point P? (Compare with I5.)

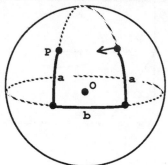

   We will study this situation
   more exactly in the coming
   pages. In effect, we are being
   forced to deal with the fact
   that sphereworld is curved.

Segments and angles in sphereworld

A **segment** in sphereworld is a piece of line (great circle) connecting two points, called the **endpoints** of the segment.

An **angle** in sphereworld is made up of two segments with a common endpoint.

To measure an angle in sphereworld:

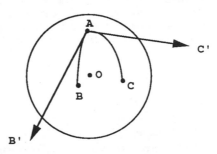

1. Find the center O of the sphere.
2. In the plane through OAB, construct the perpendicular AB' to OA through the point A. (Notice that, by exercise #6 of I25, AB' is tangent to the piece of circle AB.)
3. In the plane through OAC, construct the perpendicular AC' to OA through the point A.
4. Measure the angle B'AC'.

We measure an angle on the sphere by measuring the angle between the rays (in space) which are tangent to the sides of the sphere-angle at its vertex.

If the radius of the sphere is 1 unit, there is a very nice way to find the radian measure (I26) of angles in sphereworld:

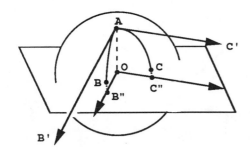

1. Construct a plane through O which is perpendicular to OA.
2. Extend the spherical segments AB and AC until they cut this new plane in points B" and C" respectively.
3. The measure of angle BAC is the same as the measure of the angle B"OC" in the new plane.

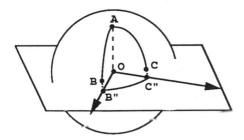

4. The circular segment B"C" is a piece of the circle of radius 1 with center O. So, by the recipe at the bottom of I26, the radian measure of the angle B"OC" is just the length of the arc B"C".

If the sphere has radius one, the radian measure of the angle BAC is the length of the arc B"C".

1. A segment between two
opposite (antipodal)
points will be called
a <u>half-circle</u>. The
region cut out on the
sphere by two half-
circles with the same
endpoints is called
a <u>lune</u>.
If the lune covers up
1/8 of the whole sphere,
what is the angle at the vertex of the
lune? Give your answer in degrees and
in radians. What if the lune covers
1/3 of the sphere? Or 1/2 the sphere?

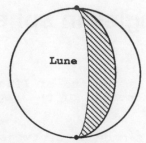

Lune

2. If the angle of a lune is α
radians, what fraction of
the sphere does it cover?

3. A triangle in sphereworld
is made up of three segments
so that each vertex of each
of the segments is a vertex
of exactly one of the other
two segments. On the sphere
at the right, draw a triangle
which has each of its three
angles equal to $\pi/2$ radians (90°).

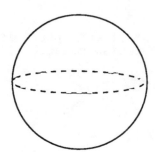

4. On the sphere of radius one
at the right, draw a spherical
triangle whose angles are $\pi/2$,
$\pi/2$, and α radians, where

    $0 < \alpha < \pi/2$.

Notice that the sum of the
(interior) angles of your
triangle is $\pi + \alpha$ radians
or $180 + (\alpha/2\pi)360$ degrees!
Compare this with I8--rules
in curved space are different!

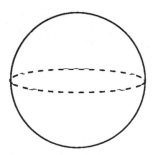

Of boxes, cylinders, and spheres

Start: Volume of 1x1x1 cube
                              is one cubic unit.

1

1

1

Volume of box
  = a·b·c cu. units
by the Second
Magnification
Principle (I32).

b   c

a

Volume of cylinder =
  (area base)(height) = πr²h

Base of the
cylinder is a
circle of radius r.

h

r

Reason: Cut the base circle up (approximately) into
        little squares. Make a box of height h over
        each little square.

  Volume = (volume of boxes) + (little bit of volume left over)
     = sum of volumes of boxes   (approximately)
     = (area 1st square)h + (area 2nd square)h + ...   (Vol. of box formula)
     = (sum of areas of the little squares)·h   (Distributive Law)

     = (area of base circle)·h   (approximately)

Repeating this process using a finer and finer division of the base circle
into smaller and smaller squares, we get better and better approximations
to the circle and to the cylinder. So the volume formula is true within a margin
of error which we can make as small as we want. We conclude that the formula
must be exactly true.

Surface area of cylinder (side only) =
  (circumference of base)(height) = 2πrh.

Reason: The side surface of the
        cylinder is like the label
        on a can. Cut it vertically,
        take it off the cylinder,
        and lay it out flat on the
        table. It becomes a rectangle
        with base 2πr and height h.

h

2πr

Volume of sphere =
  (1/3)·r·(surface area)

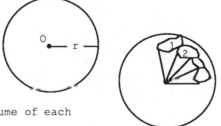

O   r

1
2

Reason: Divide the surface of the sphere up into
        small pieces. If the pieces are small
        enough, they are almost flat (like your
        back yard is on the surface of the earth.)
        Make the pyramid over each piece with vertex
        O, the center of the sphere. By I33, the volume of each
        pyramid is (1/3)(area of base)·r. So,

        volume of sphere = (sum of volumes of pyramids)   (approximately)
            = (1/3)(1st area)·r + (1/3)(2nd area)·r + ...
            = (1/3)((1st area)+(2nd area)+...)·r   (Distributive Law)
            = (1/3)(surface area of sphere)·r

So what is the surface area of a sphere of radius r ?? See I38!

1. Find the volume of a box
   with dimensions 3 widgets
   by 4 widgets by 5 widgets.
   Find the volume of a box
   with dimensions 3 astronomical units by 4 astronomical units by 5 astronomical units.

2. Find the volume of a (right
   circular) cylinder with
   height 4 meters and base
   a circle of radius 2 meters.

3. Find the (lateral) surface
   area of the cylinder in
   exercise #2. If it takes
   3 cans of paint to paint
   a square one meter on a side,
   about how many cans of paint
   will it take to paint the
   cylinder, including its top
   and bottom?

4. What is the volume of a
   sphere with surface area
   25 sq.widgets and radius
   1.4 widgets?

5. Find the surface area of
   a sphere whose volume is
   100 cu.m. and radius is
   2.88 m.

6. Suppose the volume of a
   cylinder of height 1 is
   exactly 25.9862, and the
   area of the circular base
   is exactly 25.9861. Fill
   the base with little squares
   so that the total volume of
   all the boxes over the squares
   is within 0.00005 of the
   volume of the cylinder. What
   is the contradiction?

If it takes one can of paint to paint a square one widget on a side, how many cans does it take to paint a sphere with radius r widgets?

To answer this question, begin by canning the sphere of radius one widget:

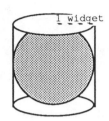

We will see that the surface area of the sphere is the <u>same</u> as the surface area of the label of the can!

By I37, the area of the label of the can is $2\pi rh = 2\pi \cdot 1 \cdot 2 = 4\pi$ sq.w. Therefore:

Surface area of sphere of radius one = 4π.

The sphere of radius r is obtained from the sphere of radius one by magnifying by a factor of r in all directions, and surface area is two-dimensional, so by the First Magnification Principle (I16):

Surface area of sphere of radius r   = r²·4π = 4πr².

So we'll have everything if we can just see why the area of the sphere of radius one is the same as the area of the label of the can. To see this, cut the sphere and can into alot of thin horizontal slices:

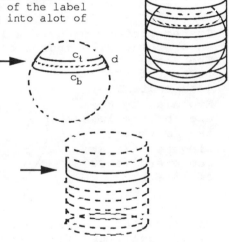

Each slice of sphere is very close to being a collar so that, by the formula on the bottom of I34, its area is very close to
(average of c_t and c_b)(slant height d) = $2\pi r \cdot d$,
where r is the radius of the dotted circle going around the middle of the collar.

Each slice of can is a cylinder so, by I37, its area is $2\pi h$.

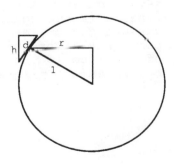

So the area of a slice of the sphere would be approximatelly equal to the area of the corresponding piece of label if we could show that
$r \cdot d = h$,
or, what is the same thing, that
$h/r = d/1$.

By I20, we can show this last equation by showing that the two triangles in the picture at the right are similar. (We magnified the picture a bit to make it easier to see.) But in exercise #7 of I20e we showed that these two triangles are indeed similar!

Each slice of sphere has about the same area as the corresponding slice of the label of the can. Adding up slices, the area of the sphere is approximately the same as the area of the label. By taking thinner and thinner slices, the margin of error in the approximation can be made as small as we want.

1. Find the surface area
    of a sphere of radius
    14 widgets.

2. Explain why the formula
    for the volume of a sphere
    of radius r is $(4/3)\pi r^3$.

3. Find the volume of the spheres
    of radius 3 widgets and of
    radius 4 meters.

4. It takes one can of paint
    to paint a square one widget
    on a side. It took 20 cans
    of paint to paint a certain
    sphere. What is the radius
    of the sphere?

5. A lead ball is dropped into
    a full swimming pool and 1
    cubic meter of water spills
    out. What is the radius of
    the ball?

6. In the picture we have a lune
    on a sphere with radius one widget.
    The angle at the
    vertex of the lune
    is α radians.
    Explain why the
    area of the lune
    is 2α square
    widgets. (Hint:
    See exercise #2
    of I36e.)

7. The formula in exercise
    #6 is really the first one
    in which we see a big
    advantage in using radian
    measure rather than degree
    measure for angles. Write
    the formula for the area
    of the lune if the angle
    has measure β <u>degrees</u>.

Excess angle formula
for spherical triangles

The sphere at the right has radius one.

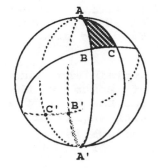

The shaded region ABC at the top of the sphere
is a spherical triangle. Its sides are
segments of great circles (lines in
sphereworld). The triangle has three
angles:
Call α the radian measure of the angle at A.
Call β the radian measure of the angle at B.
Call γ the radian measure of the angle at C.

There are two lunes in the picture above which
have vertices A and A' and angle α (one
down the front of the sphere, the other
down the back).

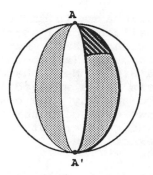

In the same way, there are two lunes
in the picture above which have vertices
B and B' and angle β.

Also, there are two lunes in the
picture above which have vertices C
and C' and angle γ.

By exercise #6 of I38e, the sum of the areas of all six lunes is

$$2\alpha + 2\alpha + 2\beta + 2\beta + 2\gamma + 2\gamma = 4(\alpha + \beta + \gamma).$$

Now look back at the top picture again. The six lunes we listed
cover up the entire sphere—they cover the spherical triangles
ABC and A'B'C' each three times and cover up the rest of the sphere
exactly once. Therefore
(sum of areas of six lunes listed above) =
    (area of sphere) + 2(area triangle ABC) + 2(area triangle A'B'C').

Since A', B', and C' are the opposite points of A, B, and C respectively,
(area triangle ABC) = (area triangle A'B'C'), and so we get
        $4(\alpha + \beta + \gamma) = 4\pi + 4$(area triangle ABC)
  or
        (area triangle ABC) $= (\alpha + \beta + \gamma) - \pi$.

**Area of spherical triangle on sphere of radius
one unit is:
 (sum of radian measures of its interior angles) − π
square units.**

1. A triangle on a sphere
   of radius one has all
   its angles of measure
   $2\pi/3$ radians. Find its
   area.

2. A triangle on a sphere
   of radius one has angles
   of 90°, 70°, and 60°.
   Find its area.

3. Explain why, on a sphere
   of radius r, the formula
   for the area of a triangle
   with angles of α, β, and γ
   radians is
   $$r^2 \cdot (\alpha + \beta + \gamma - \pi).$$

4. With a magic marker and a
   balloon, draw a spherical
   triangle ABC. Put in the
   opposite (antipodal) triangle
   A'B'C'. Next shade in the
   six lunes we used in I39
   to see why they cover the
   two triangles three times
   and the rest of the sphere
   exactly once.

5. Find a formula for the area
   of a spherical quadrilateral
   in terms of the radian
   measures of its four angles.
   (Hint: Divide the quadrilateral into two triangles.)

6. Find a formula for the area
   of a spherical pentagon
   in terms of the radian
   measures of its five angles.

7. Find a formula for the area
   of a spherical n-gon
   in terms of the radian
   measures of its n angles.

Hyperbolic-land

Besides the plane and the surface of the sphere, there is one
more two-dimensional geometry we want to talk about:

The universe is the <u>inside</u> of a circle,
but the distance d(P,Q) between two points P and Q is weird, in fact
you need to know about logarithms and cross-ratio (I31e) even to
write down the formula.

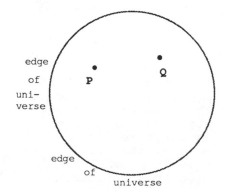

To find any point R on the "straight line" segment
between P and Q, we ask the same question we do
in ordinary geometry, or in spherical geometry:

Which are the points R that lie on the shortest
path from P to Q?

or, which is the same thing:

Which are the points R so that
$$d(P,R) + d(R,Q) = d(P,Q) ?$$

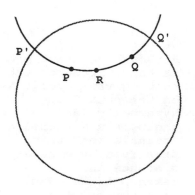

The answer turns out to be that the shortest
path from P to Q in hyperbolic-land is along
a circular arc! In fact the circle must be the one
through P and Q which meets the edge of hyperbolic-
land perpendicularly at P' and Q'. (We see in C23 how
to construct that circle—also see page CP22.)

We should also say that the edge of hyperbolic-land
is "infinitely far away," that is, as P moves away
from Q toward the edge of hyperbolic-land, d(P,Q)
gets bigger and bigger without bound. You can
never get to the edge of hyperbolic-land because
you would have to travel an infinite (hyperbolic)
distance to get there! The formula for d(P,Q) turns
out to be
$$d(P,Q) = (1/2)|\log (\text{cross-ratio } (P,P',Q,Q'))|.$$
(See exercise #10 of I31e.)

All the properties we listed in I1 for lines are true
for hyperbolic lines.

**Lines in hyperbolic-land, which we call
<u>hyperbolic lines</u>, are circular arcs
which meet the edge of hyperbolic-land
perpendicularly.**

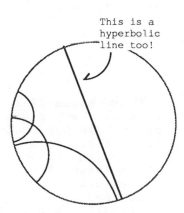

This is a
hyperbolic
line too!

All the arcs at the right are hyperbolic lines.
The closer the hyperbolic line gets to going
through the "center" of the circle which defines
the edge of hyperbolic-land, the more and more
the hyperbolic line becomes like an ordinary
line segment in the ordinary plane.

1. If you know a little
    about logarithms,
    explain why
    d(P,Q) goes to
    infinity as P
    moves toward the
    edge of the
    universe, that
    is, toward P':

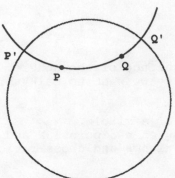

2. Show that rotation of
    hyperbolic-land around
    the "center" O of
    our disc-universe
    is a hyperbolic
    motion, that is,
    that it preserves
    hyperbolic distance:
    (Hint: Use P7.)

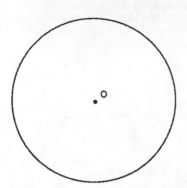

3. Although we can't prove
    it here, it turns out
    that, given a point X
    on the path C in
    hyperbolic-land,
    there is a hyperbolic
    "translation" which
    preserves hyperbolic
    distance and
    takes O to X. Use
    this to show that,
    given any points P
    and Q in hyperbolic-
    land, there is a
    hyperbolic motion
    taking P to Q.

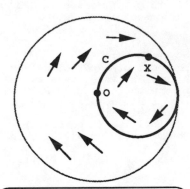

action of hyperbolic
translation: everybody inside
the universe moves but stays
inside the universe

4. Use exercises #2 and #3
    to show that, given any
    point P in hyperbolic-
    land, there is a
    hyperbolic motion
    which leaves P
    where it is and
    "rotates"
    hyperbolic-land through
    any given angle around P.

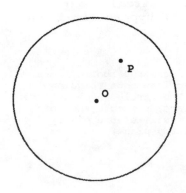

5. Sketch a triangle in hyperbolic-land. Remember that its sides must be hyperbolic lines, that is, pieces of circles, and that these circles must hit the edge of hyperbolic-land perpendicularly.

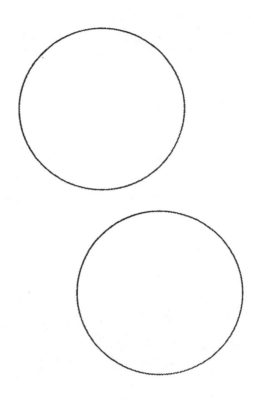

6. Do the interior angles in a hyperbolic triangles add up to 180˚? More than 180˚? Less than 180˚? To answer this, draw a hyperbolic triangle which has two sides on "ordinary" line segments through the "center" of hyperbola-land. (By exercises #3 and #4, you can move any hyperbolic triangle into this position without hyperbolically bending, stretching, or tearing it.)

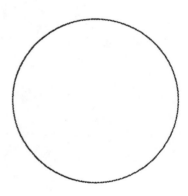

7. In ordinary geometry, two lines are parallel if they don't intersect (I14), and, through a point not on a line, there passes one and only one parallel line (I12e, ex.#4). Draw a picture to show that in hyperbola-land, at least one of these two "facts" must be wrong for hyperbolic lines.

8. You may be tempted to define parallel lines in hyperbolic-land as we did in I11, namely, as extensions of opposite sides of a rectangle. Use ex. #6 to show that there are no rectangles in hyperbolic-land!

Construction

Copying triangles

Make sure you know how to copy a segment
(see I4e, exercise #1) before you do this page.

Copy this:

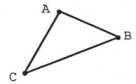

Put a point wherever you want, and
put a line through it. This point
will be the copy of point A, and
the copy of side AB will lie on
the line.

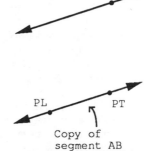

Use your compass to copy segment
AB onto the line so that A goes
to the given point.

Remember:

PT = **The point of the compass**

PL = **The pencil of the compass**

Set your compass to the
length of segment AC on
the given triangle.
Without changing the
setting, place the point
of the compass on the
copy of point A, and
make a circle.

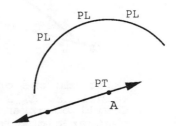

Set your compass to the length
of segment BC on the given
triangle. Without changing
the setting, place the point
of the compass on the copy of
point B, and make a circle.

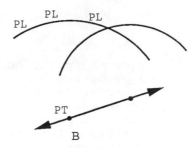

Where the two circles cross
is where to put the copy of
C. Now, you can use your
straight edge to connect
A to C, and B to C. The
triangle you have now is an
exact copy of the original
triangle ABC. This triangle,
we say, is congruent to the
original. You can test this with tracing paper.

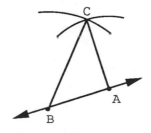

1. In the space at the right, copy the following triangle:

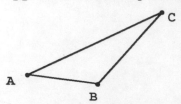

2. Copy the following triangle in the space to the right. The copy of segment AB is already given, but you must copy the triangle in two different ways so that segment AB remains unmoved.

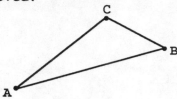

(Hint: In one of the solutions, A, B, and C will be arranged in counter-clockwise order, in the other, in clockwise order)

3. Once the position of the copy of segment AB in exercise #2 was determined, there were only two possible positions for the copy of the vertex C. Explain why.

4. Notice that the only
    information we need to
    contruct a triangle is
    the length of the three
    sides. (That was all we
    used in C1.) Using only
    straight-edge and compass,
    construct a triangle with
    sides given below:

    _____2_____

    _____3_____

    _____4_____

5. Construct a triangle with
    sides 2, 3, and 3.

6. Try to construct a triangle
    with the three sides given
    below:

    _____2_____

    _____2_____

    _____5_____

    Explain what goes wrong.
    (Guess a rule.)

Copying angles

Copy this:

Open compass to a
convenient setting.
Put the point on the
vertex of the angle, and
draw part of a circle
with the pencil. <u>Leave
your compass open to this
setting</u>.

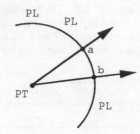

Remember:

PT   PL

PT = The point of the compass

PL = The pencil of the compass

Forget the first step for a moment.
Pick any point you want to be
the vertex of the copy of
the angle. Draw whatever
ray you want from that
point. It will be one
of the two rays of the
copy of the angle. Now,
go back to the first
step. Pick up your
compass opened to the same
seting, and draw a circle
around the copy of the vertex.

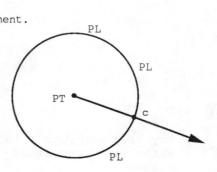

Open your compass to the
distance between the points
a and b   in the first
step. Place point of
compass on point c
constructed in step two.
Draw a circle.

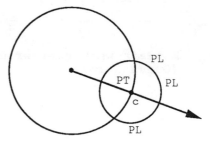

Pick either point at which the two circles
intersect, and draw the ray from the vertex
through that point. These two rays
give an exact copy of the original
angle. Check it with your tracing
paper.

1. **In the space at the right, copy the following angles:**

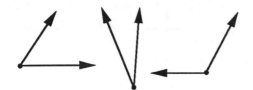

2. **Use your straight edge and compass to "add" the three angles given in exercise #1. (Remember: to "add" two angles you copy them so that they have the same vertex and a common side, and so that their insides do not overlap.)**

Constructing perpendiculars

Two lines are called **perpendicular** if they
intersect in a point O so that a ray from
O along one of the lines and a ray from O
along the other make an angle of 90˚.

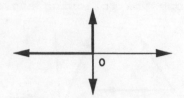

PT = **The point of the compass**

PL = **The pencil of the compass**

Through a given point on a given line,
construct a line perpendicular to the
given line.

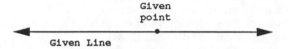

Open the compass to a convenient setting.
Place the point of the compass on the
given point O, and make a circle. Mark
the two points A and B where the
circle meets the given line. The important
thing is that the distance between O
and A is the same as the distance
between O and B.

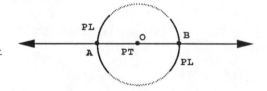

Open the compass to a somewhat
larger setting. Put the point of the
compass on A and make a circle.
Without changing the setting of the
compass, put the point on B and
make another circle. Mark the two
points C and C' where the two
circles intersect.

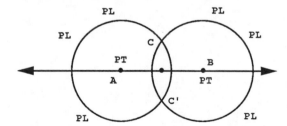

Remember that "≈" means "is congruent to".

Use the straight edge to draw the ray
from O through C. Now, OA ≈ OB,
and AC ≈ BC, by the way we've
constructed things. Also, OC ≈ OC.
We'll see in I10, that this
implies (triangle AOC) ≈ (triangle BOC).

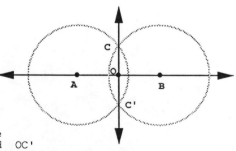

So, ∠ AOC ≈ ∠ BOC. But together, these
two angles make up an angle of 180˚. So,
each must have a measure of 90˚. In the same
way, if we draw the ray OC', it makes an angle
of 90˚ with the ray OA. So, the rays OC and OC'
fit together to make a line perpendicular to the given line.

1. **There is a construction which is almost the same as the construction:**

   Through a given point on a given line, construct a line perpendicular to the given line.

   It is the construction:

   Through a given point <u>not</u> on a given line, construct a line perpendicular to the given line.

   In fact, both constructions start with a circle around the given point in order to contruct two points, A and B, on the given line. Thereafter, the two constructions are identical. In the space at the right of this exercise, make the second construction.

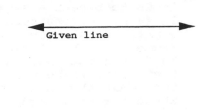

2. **Use your compass and straight edge to construct perpendiculars to the given lines through the given points:**

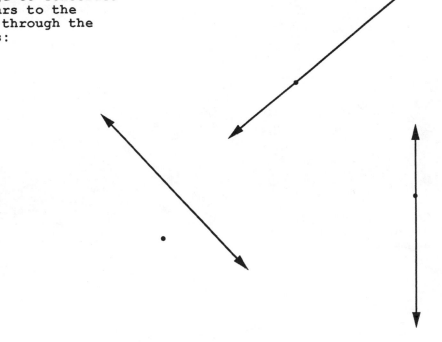

3. You can use a shortened
   version of the construction
   in C3 to bisect a segment,
   in fact, to construct a
   a perpendicular to the
   segment which cuts it in
   half. In the space at
   the right, write the
   text that should accompany
   the diagrams in order to
   explain how to construct
   the <u>perpendicular</u> <u>bisector</u>
   to the segment AB.

4. Construct the perpendicular
   bisectors of the given
   segments.

5. After you have done I18,
    come back and explain
    why the construction in
    exercise #3 really does
    make a perpendicular
    which bisects the
    original segment.
    Namely, show that

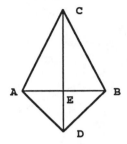

    if d(A,C) = d(B,C) and
    d(A,D) = d(B,D), then
    ΔADC ≈ ΔBDC.
    Use this to show that
    ΔAEC ≈ ΔBEC. This
    in turn means that
    d(A,E) = d(B,E) and
    that CD makes a 90°
    angle with AB.   Why?

Constructing parallels

**Through a point P not on a given line L, construct the
line parallel to the line L.**

Line L and point P are shown at the right.

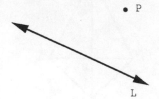

Through P, draw any line M that intersects
the line L at some point. The angle that
L makes with M at the point of intersection
wil be called ∠ α.

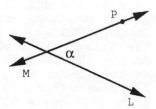

Using the instructions in C2, copy ∠ α so that
the copy has vertex P, and one ray of the angle
lies along M (pointing the same direction as
it did originally).

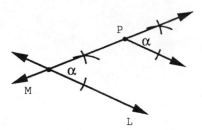

Complete the other ray of the copy of ∠ α to
obtain a line parallel to line L, and passing
through the point P. This new line and L are
parallel by the Principle of Corresponding
Angles (see I14e).

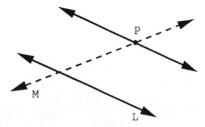

1. Using only your straight
   edge and compass, construct
   parallels to the given
   lines through the given
   points.

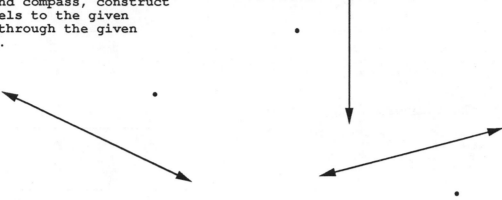

2. A "parallelogram" is a
   quadrilateral whose
   opposite sides are
   parallel:

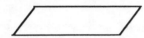

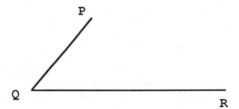

   Using only your
   straight edge and
   compass, complete the
   figure at the right
   to a parallelogram
   such that the given
   figure forms two side
   of the parallelogram.

3. Can you construct a line
   through a point P which
   is parallel to a given
   line L, when P lies on
   L?

Constructing numbers as lengths

We pick any length we want to represent the number 1:   ●———●

Question: Starting from the length which represents the number 1, which lengths can
we construct using only straight edge and compass? Which numbers do these lengths
represent? (You will need to do I20 to understand the theory behind this page.)

Whole numbers:

To construct any whole number, first
draw a horizontal ray. Set your compass
to one unit, place the point of the
compass on the endpoint of the ray, and
mark off one unit from the endpoint.
Placing your compass point at the place
where the mark intersects the ray, make
another mark, as shown in the second
picture at the right. Continue this
until you have reached the number you want
to construct. In this case, the distance
from point A to point B is 5 units.

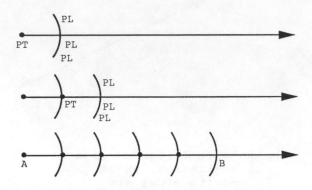

Addition:

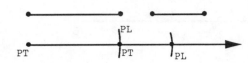

To construct the sum of two given lengths,
draw a ray and mark the first length off,
starting from the endpoint of the ray. Then
mark the second length in the same direction,
starting from the endpoint of the first.

Subtraction:

To construct the difference of two given
lengths, draw a ray and mark the first length off,
starting from the endpoint of the ray. Then
mark the second length in the opposite direction,
starting from the endpoint of the first.

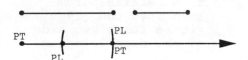

Multiplication and Division:

Construction tool: Suppose you are given

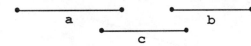

Lay them out on two rays of any old angle
as shown at left. Draw the segment MN.

Any old angle
you want

Then construct the line through Q which
is parallel to MN (see C4). Since MN and
PQ are parallel, then by I21: b/a = e/c.
So if c = 1, then e = b/a.
Likewise, if a = 1, then e = b·c.

Now we are ready to multiply and divide:
Say the two given lengths you want to
multiply or divide are x and y.
To multiply x and y,
set b = x, c = y, and a = 1.
Then e = x·y.
To divide x by y,
set b = x, a = y, and c = 1.
Then e = x/y.

•——• Use this as the length for 1 unit.

1. Construct the number 2, and construct the number 7. Note that with a given length of 1, you can construct any whole number. (If you can't fit some of the constructions on this page, do them on another sheet and attatch it to this one).

2. Construct the fraction 1/2 , and construct the fraction 1/3. Note that with a given length of one, you can construct any fraction.

3. Construct the product 4/3 × 3. (Hint: You can construct 4/3 by adding 1 and 1/3.)

4. Explain why
   $d(A,E) = p/q$:

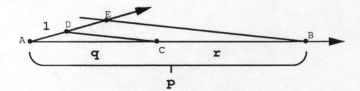

Given a number, construct its square root

•———•   This will serve as the length of one unit.

On this page, we will construct √3.

First, construct the number 3, of which you wish to find the square root (See C5).

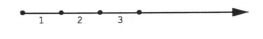

Next, add one to this number, by constructing the sum 3 + 1 (See C5).

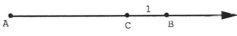

Now, bisect the segment AB (see ex. #3, of C3e). Where the bisector intersects the segment is the exact center of the segment, called the **midpoint**. Call this midpoint M.   d(A,M) = d (M,B)

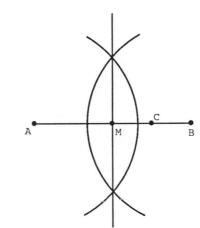

Set your compass to the distance between A and M, and draw a circle with center M and radius d(A,M). Now draw a perpendicular to AB through C (see C3). Where this perpendicular intersects the circle, call it point E. Lastly, draw segment AE and segment EB.

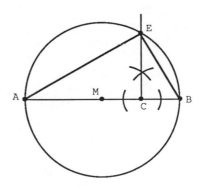

By ex.#12 of I20: $\dfrac{d(B,C)}{d(E,C)} = \dfrac{d(E,C)}{d(A,C)}$

Since: $d(B,C) = 1$

Then: $\dfrac{1}{d(E,C)} = \dfrac{d(E,C)}{d(A,C)}$

So: $d(E,C)^2 = d(A,C)$

And: $d(E,C) = \sqrt{d(A,C)}$

So: $d(E,C) = \sqrt{3}$

Note: By combining constructions of C5 and C6, we can now construct numbers as diverse as $\frac{1}{2} + \frac{1}{3}\sqrt{3} - \frac{1}{4}\sqrt{7}$ and $1/(1-\sqrt{2})$. There are, however, some numbers that are impossible to construct. Over 150 years ago a French mathematician named E. Galois showed that, for example, it is impossible to construct the cube root of two with straight edge and compass. This means we can't construct the edge of a cube whose volume is two cubic units.

•————•   **Use this as the length for 1 unit.**

1. Construct $1 + \sqrt{2}$.

2. Construct $\sqrt{1 + \sqrt{3}}$.

3. Construct $\dfrac{1}{\sqrt{1 + \sqrt{3}}}$.

4. Construct $\sqrt{5} - 1$. This number is a very important one, as we will see in C9. One-half of this number is called the "golden ratio":

Use the quadratic formula to show that, if the two rectangles in the picture are proportional, that is, if

$$x/1 = 1/(1+x),$$

then $x = (\sqrt{5}-1)/2$.

Constructing parallelograms

Be sure that you have read I7e before starting this page, and understand these principles of parallelograms: Opposite angles are equal, and opposite sides are congruent. Also make sure you know how to copy angles (see C2).

We must construct a parallelogram, using only the following information: Given an angle and two segments, construct the parallelogram with one of its four angles congruent to the given angle, and the two sides adjacent to the angle equal in length to the two given segments:

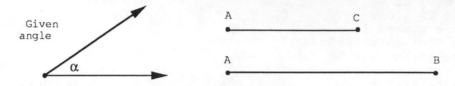

First, draw a horizontal ray. Copy the given angle onto the ray so that the vertex of the angle is the endpoint of the ray. Then, take the distance of the first given segment, and copy it onto one of the two rays (it doesn't matter which one). Copy the other given segment onto the ather ray of the angle. Make sure one endpoint of each of the copied segments is A, the vertex of the angle.

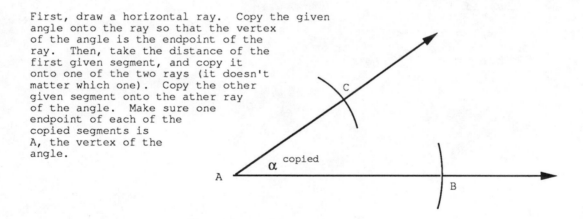

Now, we must find the fourth vertex (D) of the parallelogram to complete the figure. Since AB ≈ CD and AC ≈ BD then the point D must be d(A,B) distance from C and d(A,C) distance from B. Set your compass to d(A,B), put the point on C, and make a good-sized arc about where you think D will be. Then set the compass to d(A,C), put the point on B, and make another arc. Where the two arcs intersect is point D. Draw segments CD and BD, and the parallelogram is complete.

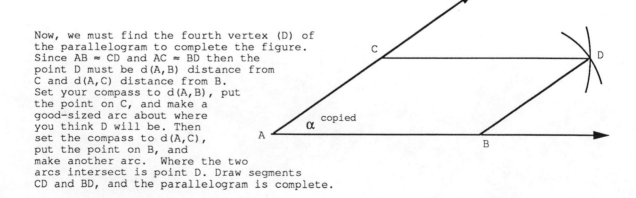

1. Construct the parallelogram
   so that two opposite angles
   of the four angles are
   congruent to the given
   angle (below), and so that
   each of the two pairs of
   opposite sides are
   congruent, respectively,
   to the two given segments.

2. All of the angles in
   a parallelogram add up
   to 360°. Opposite angles
   of a parallelogram are
   congruent. So, once we know
   one of the angles (like
   angle α in exercise #1) we
   can solve for the other
   angles:

   $2\alpha + 2\beta = 360°$

   $\beta = 180° - \alpha$

   So, if we can determine all
   four angles of the
   parallelogram when given
   just one, why can't we
   construct a unique
   parallelogram given
   just one angle?
   Give examples.

Constructing a regular 3-gon and 4-gon

A regular 3-gon is an "equilateral triangle" and a regular 4-gon is a square.

Draw a horizontal segment, making it rather large to give yourself planty of working space. Mark any point on this segment, and call it A. Open the compass to a certain distance (any distance will do for now), place the compass point on A, and mark off a point on the segment on either side of A. Call this point B. Now, keep the same compass setting, keep the point on A, and draw an arc above the segment about halfway between A and B. Now put the point on B, and do the same thing with the same compass setting. Where the two arcs that you have drawn intersect, mark point C. This point is the same distance from A as it is from B, and to complete the triangle just draw segemnts AC and BC. All three sides should be congruent to one another.

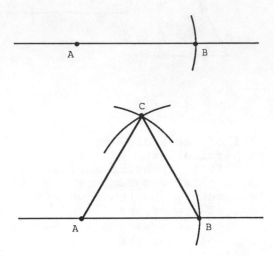

Make sure you know how to construct perpendiculars (see C3) before trying to construct a square. First, draw a horizontal segment and mark a point A on it, much like above, when constructing an equilateral triangle. Now, construct a perpendicular to the segment at A. Now, for the length of the sides of the square, draw a segment, off to the side of the drawing, to represent the length of each side, so that you can set your compass to this length when necessary. Call its length s. Next, mark off length s along the base segment from A (call the endpoint B), and mark length s up along the perpendicular from A (call the endpoint C). Keeping the compass set to s, make an arc above B (with compass pt. on B), and to the right of C (with the pt. on C). These two arcs should intersect to the upper right of A. Call this point D. Draw segment BD and CD. This completes the square (shown in bold for clarity).

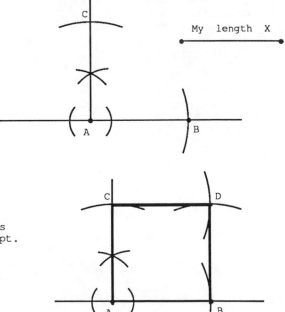

1. Instead of using a random length for the side of an equilateral triangle, you could use a previously determined length. Construct the equilateral triangle with a side of:

2. Construct a "house" (see below) so that each side of the square and each side of the equilateral triangle are the same as the given length:

    The "house" should look like this:

3. We constructed a regular 4-gon on C8. Part of the definition of a "regular" polygon is that all of the angles in it are congruent. Since each angle of a square is 90° and perpendicular lines form a 90° angle, what would be another way to construct a square just by making a succession of perpendiculars?

    (Hint: You would use principles outlined in I4e, ex.#1, and C3.)

Constructing a regular 5-gon

We must start by designating a certain length as "one unit." In this case <u>it</u> is 2 centimeters. After that we must determine the length of $\sqrt{5} - 1$ units (see C6e).

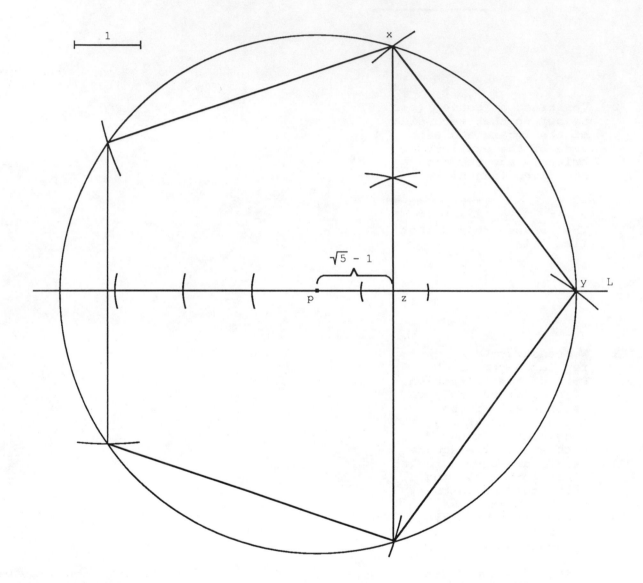

Start with a horizontal line, L. Pick a point p on L that will be the center of the pentagon. At a distance of $\sqrt{5} - 1$ units from p along L, mark point z. Construct a perpendicular to L at z. Next, construct a circle of radius 4 units about p. The distance from point y (the point where the circle intersects L at the right) to point x (the point where the circle intersects the perpendicular to L at the top) is the length of one side of the pentagon. Place the compass point on x and mark off this distance on the circle. Then place the compass point on the point you just marked, and mark off the distance again. Repeat this five times, and then connect all of the points you have marked. This is the regular 5-gon, or pentagon.

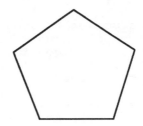

1. Given a regular pentagon
    find the center of the
    circle which passes through
    the vertices of the pentagon.
    This cirle is called the
    <u>circumscribed</u> circle, and its
    center is called the <u>center</u>
    of the polygon. (Hint: Con-
    struct the perpendicular
    bisectors (C3e) of two sides
    of the pentagon. The bisectors
    will intersect in the center
    of the pentagon.)

2. Explain why the center of a
    circle lies on the perpendi-
    cular bisector of any chord,
    that is, explain
    why, if a = b
    and $\alpha = 90°$,
    then r = s.

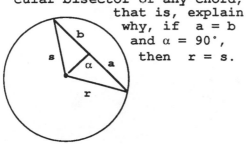

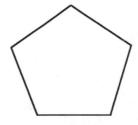

3. An <u>apothem</u> of a regular
    pentagon is is the segment
    from the center to a side
    of the pentagon which is
    perpendicular to the side.
    Construct all five apothems
    to the pentagon at the
    right.

4. A circle inside a regular
    pentagon is said to be
    inscribed if it just "touches"
    (is <u>tangent</u> to) each side of
    the pentagon. Construct the
    circle inscribed in the
    pentagon in exercise #3.
    (See I25 for the definition
    of "tangent.")

5. Our construction for the
   pentagon on C9 is used to
   construct a pentagon inside
   a given circle. A different
   construction gives a pentagon
   with sides of given length:
   First take a side, like the
   one below. Call it 1 unit.

   Now use this distance to
   construct the distance
   $(1+\sqrt{5})/2$ by the methods on
   pages C5 and C6. Use this
   length to construct a
   triangle ABC where AB =
   $(1+\sqrt{5})/2$, BC = 1, and AC = 1.
   You can use the method on
   "more C1e" to construct the
   triangle given these three
   sides. Below is the distance
   of $(1+\sqrt{5})/2$, according to the
   given distance of 1, above.

$$\frac{1+\sqrt{5}}{2}$$

   Here is what the
   constructed triangle
   looks like:

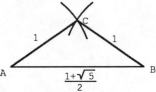

   Sides AC and CB are two
   sides of the pentagon we
   want. By copying this
   triangle (building on
   BC as in diagram below)
   by the method on C1, we
   can complete the pentagon.

   Using the
   segment
   at the
   top right
   of the
   page as
   one unit,
   construct a
   regular pentagon
   with side one.

   You may need another piece
   of paper.

Use this as one unit
for <u>your</u> construction.

6. To understand the mathematics
   behind the construction of the
   regular pentagon, we will need
   some trigonometry, namely we
   need the formula for the cosine
   of the sum of two angles:

   $\cos(\alpha + \beta) = \sin(90° - (\alpha + \beta))$

   $\qquad = \sin((90° - \alpha) - \beta)$

   $= \sin(90° - \alpha) \cdot \cos\beta - \cos(90° - \alpha) \cdot \sin\beta$

   $= \cos\alpha \cdot \cos\beta - \sin\alpha \cdot \sin\beta$

   Where are the formulas in I29e and
   I30e which justify this calculation?

7. Show that

   $\cos 2\alpha = \cos^2\alpha - \sin^2\alpha$

   $\qquad = \cos^2\alpha - (1 - \cos^2\alpha)$

   $\qquad = 2\cos^2\alpha - 1.$

8. Explain why
   $\cos(2 \cdot 72°) = \cos 144° = -\cos 36°.$

   Explain why
   $2[\cos(2 \cdot 72°)]^2 = 2[\cos 36°]^2$

   $\qquad\qquad\qquad = \cos 72° + 1.$

   Explain why
   $2[2(\cos 72°)^2 - 1]^2 = \cos 72° + 1.$

   So, if we let X = 2 cos 72°,
   show that
   $X^4 - 4X^2 - X + 2 = 0.$

   Factoring we get
   $(X^2 - X - 2)(X^2 + X - 1) = 0.$

   Why can't $X^2 - X - 2 = 0$?

   Conclude that
   $X^2 + X - 1 = 0$
   and solve for X using the quadratic
   formula.
   Where have we seen this number
   before?

9. Explain what exercise #8 has to do
   with the construction of the regular
   pentagon.

Constructing a regular 6-gon

First, designate a certain length as "one unit." Then, decide how
long, in units, you want a side of the regular 6-gon (hexagon) to be.
Call this length "s." Here, s = 4.

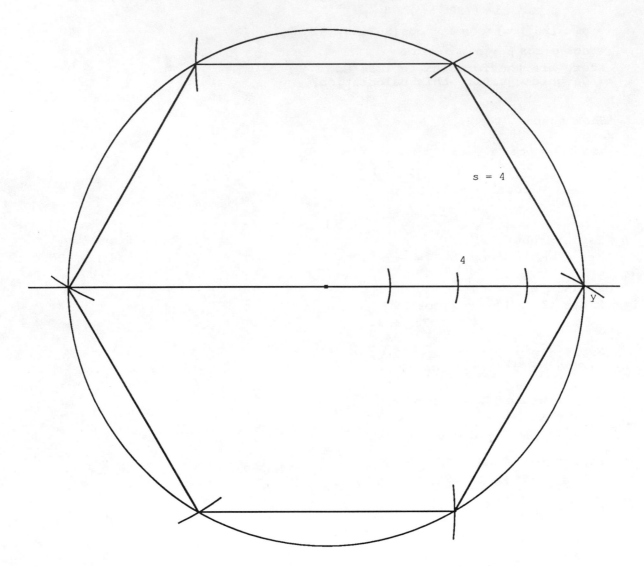

Draw a horizontal line, select a point on it to be the center of your hexagon,
and construct a circle of radius s about this point. Now, mark off distance
s on the circle from point y (the point where the circle intersects the
right side of the horizontal line you drew first). Now, mark off distance s
on the circle from the point you just marked, and repeat this process six times.
When you connect all of the points you marked, you will have the hexagon.

1. In the space below, construct
   the hexagon with a side of 2
   units, using the following
   distance as one unit:

   1
   ├──────────────┤

2. Explain why each side of a
   hexagon is equal in length to
   the radius of the circle
   used to construct it.
   (Hint: Think of a
   hexagon as being made up of
   six congruent equilateral
   triangles.)

3. Construct the center of
   the hexagon at the right.
   Also construct the circumscribed circle, at least
   two apothems, and the
   inscribed circle. (See
   C9e.)

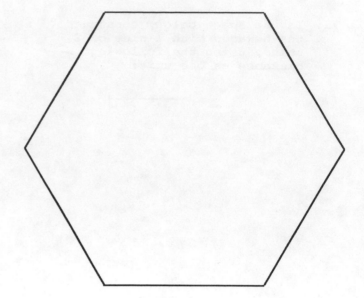

4. For any regular polygon,
   there is a relation between
   the radius r of the circumscribed circle, the radius a
   of the inscribed circle
   (apothem), and the length s
   of a side of the regular
   polygon. Look at the picture
   at the right and write down
   the formula.

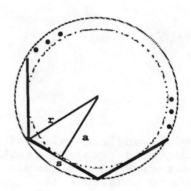

Constructing a regular 7-gon (almost)

No one has ever been able to give a way to construct a perfect regular 7-gon
using only straight-edge and compass, as we have just done for 3-,4-,5-, and 6-gons.
In the 19th century, mathematicians E. Galois and K.F. Gauss discovered why.
To be constructible with ruler and compass, the lengths involved have to be
gotten by solving a chain of second-degree equations starting from rational numbers.

The 7-gon turns out to be impossible because the prime number 7 cannot be written in
the form $2^m + 1$!   (Notice that $3 = 2^1 + 1$, $5 = 2^2 + 1$, $17 = 2^4 + 1$.)

An eighth-grade boy named Matthew Reynolds discovered the following construction
on his own, when challenged by his geometry teacher to find a "ruler and
compass" construction for a regular 7-gon.   Although it has been proven that
it is impossible to construct a **perfect** regular 7-gon, this one is an
extremely close approximation:

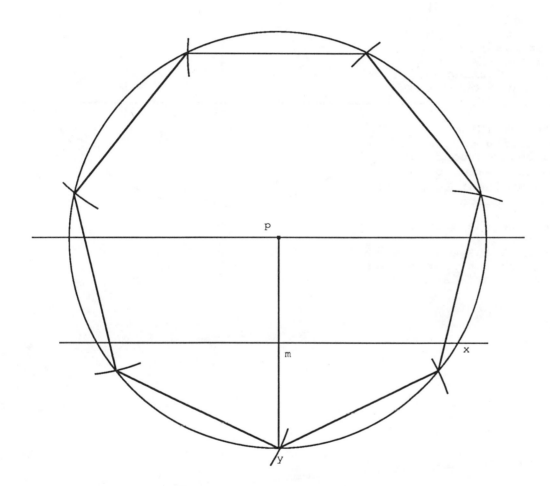

First, draw a horizontal segment.   Select a point "p" on it to be the center of
your regular 7-gon.   Construct a circle about p, of any radius. Now, construct
a perpendicular to the horizontal line at p.   Where this perpendicular intersects
the circle at the bottom, mark point y.   Construct the perpendicular bisector
of segment py. Where this intersects the circle on one side, mark point x.   Denote
by m the midpoint of segment py.   Starting at y, mark off distance
d(x,m) on the circle seven times, just like you did on C9 and C10.   Connect all of
these points, and the almost regular 7-gon is complete!   In the exercises, we
will see just how close to perfect this 7-gon is.

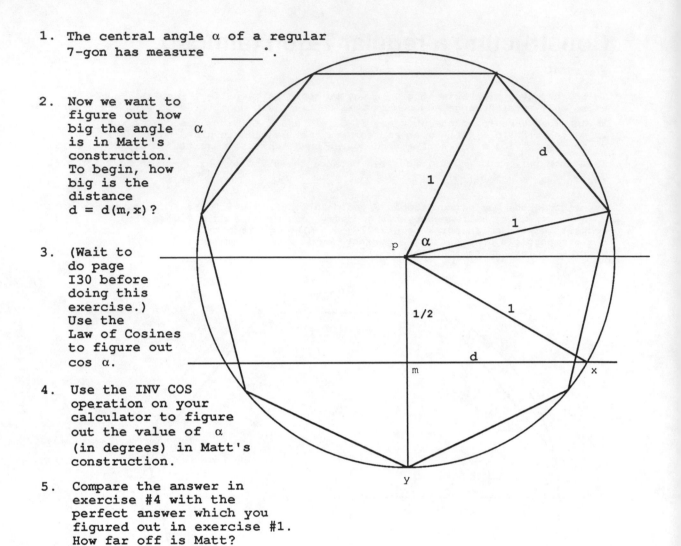

1. The central angle α of a regular
    7-gon has measure _____°.

2. Now we want to
    figure out how
    big the angle   α
    is in Matt's
    construction.
    To begin, how
    big is the
    distance
    d = d(m,x)?

3. (Wait to
    do page
    130 before
    doing this
    exercise.)
    Use the
    Law of Cosines
    to figure out
    cos α.

4. Use the INV COS
    operation on your
    calculator to figure
    out the value of α
    (in degrees) in Matt's
    construction.

5. Compare the answer in
    exercise #4 with the
    perfect answer which you
    figured out in exercise #1.
    How far off is Matt?

Constructing a regular tetrahedron

A tetrahedron is a three-dimensional, four sided figure, each of whose sides is an equilateral triangle. In a tetrahedron, each side of each triangle touches exactly one side of one other triangular face.

To construct a tetrahedron, first start with a flat figure, that can later be cut out and folded into the tetrahedron. The flat figure consists of four equilateral triangles, grouped so that one triangle is in the middle of the other three and each of the other three has one side in common with the central one. To construct these triangles, use the method on C8 to make the first one, and then construct the rest by using each side of the first one as a new baseline instead of making a new one each time.

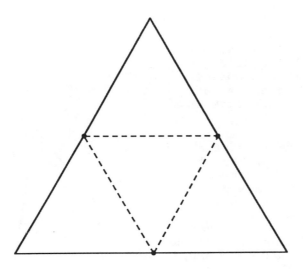

To construct the actual tetrahedron, cut out the figure you just constructed along the lines that are solid, above. Fold upward along the dotted lines, above. When all three of the outer tips of the outer triangles converge, tape the cracks closed. The complete tetrahedron should look like this:

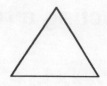

1. Discuss the symmetries of an
   equilateral triangle, that is,
   ways of picking the triangle
   up and putting it back down
   so that it occupies the same
   space. Show that you can
   rotate the triangle into itself
   by rotating in the plane by
   120° and by 240° about the
   center of the triangle. What
   happens when you rotate by
   360°?
   Show that you can get 3 more
   symmetries if you are able to
   pick up the triangle in
   three-dimensional space and
   turn it over and put it back
   down.
   (Hint: To keep track of
   different symmetries, number
   the vertices of the triangle
   before you start and notice
   where the numbers are when
   you put the triangle back down
   in the same space.)

2. Discuss the symmetries of a
   regular tetrahedron, that is,
   ways of picking the tetrahedron
   up and putting it back down
   so that it occupies the same
   space. Show that you can
   rotate the tetrahedron into
   itself by rotations of 120° and
   of 240° about an axis through
   a vertex and the center of the
   opposite face. What happens
   when you rotate by 360°?
   Show that you can rotate the
   tetrahedron into itself by
   rotations of 180° about an axis
   through the midpoint of an edge
   and the midpoint of the
   opposite edge. What happens
   when you rotate by 360°?
   (Hint: To keep track of
   different symmetries, number
   the vertices of the triangle
   before you start and notice
   where the numbers are when
   you put the tetrahedron back
   down in the same space.)

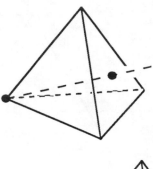

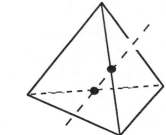

3. If you could pick up the tetra-
   hedon in four-dimensional space and
   turn it over and put it back down
   again, you could get 12 more
   symmetries.

Constructing a cube and an octohedron

To construct a cube, first construct
its six faces on a flat surface, in
the configuration at right. Then,
cut along the lines that are soild
in the diagram at right. Then fold,
along the dotted lines, the sides
marked y so that they are
perpendicular to the side marked x.
Then fold the side marked z over
the top and tape all cracks closed.

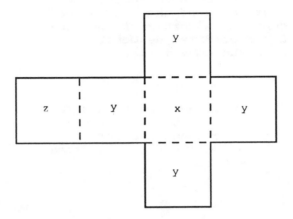

An octohedron is an eight-sided
figure whose sides are
equilateral triangles. Picture
two pyramids, each with four
faces and a base, joined at
the bases. To construct this,
Construct a square and eight
equilateral triangles in the
configuration at right. Note
that the square is not a face
and will end up inside the
octohedron. Cut out the figure
along the "solid" lines at right.
Fold the triangles inward along
the dotted lines on the left
side, and outward on the
right side, until there is a
pyramid on each side. Tape
each one closed, then fold
the lefthand one inward along
the dotted left side of the
square until the pyramid sits
on the square. Repeat this
process on the right side,
folding it away instead of
toward you. Tape all cracks
closed, and the octohedron
is complete.

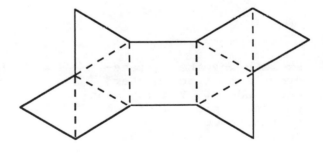

1. We say that the cube and
   the octahedron are "dual"
   to each other. This means
   that you can put a cube
   inside an octahedron so
   that ech vertex of the cube
   is the center of exactly one
   face of the octahedron, and
   every face of the octahedron
   contains a vertex of the cube.
   Build or draw the octahedron
   with the cube inside.

2. Now put an octahedron inside
   a cube so that the vertices
   of the octahedron coincide
   with the centers of the faces
   of the cube.

3. How many faces does an
   octahedron have? How many
   edges? How many vertices?
   How about the cube? Use
   the picture or model you
   made in exercise #1 or #2
   to explain why the cube
   and the octoahedron have
   the same number of edges.

4. Explain why the tetrahedron is dual to another
   tetrahedron. You may want
   to draw a picture of two
   tetrahedrons in dual
   position to each other.

Constructing a dodecahedron and an icosahedron

A **dodecahedron** is a twelve-sided figure with regular 5-gons (pentagons) for its sides. An **icosahedron** is a twenty-sided figure with regular 3-gons (equilateral triangles) for sides.

To construct the three-dimensional shape of the dodecahedron, construct two sets of six pentagons in the configuration below (See C9). Cut out each set of six and fold along the lines where the pentagons are joined until each side of the cracks (example indicated by arrow) is joined to the other side. See how the two resulting bowl-shaped figures can fit together. Tape them together.

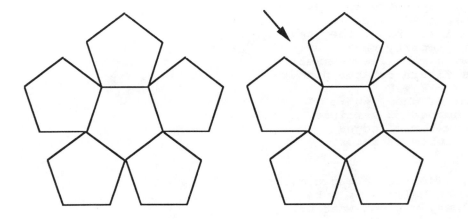

To construct the three-dimensional shape of the icosahedron, construct twenty equilateral triangles in the configuration below. Cut out the group. Bend around until line labeled x meets line labeled y. Tape these together. Then fold the topmost triangles inward until they all meet at point a, and all bottommost triangles until they meet at point b. Tape these together, and the icosahedron is complete.

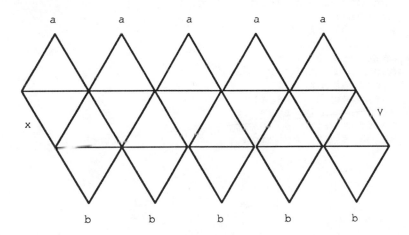

1. How many faces does a
   dodecahedron have? How
   many edges? How many
   vertices?
   ___faces, ___edges, ___vertices.
   How about an icosahedron?
   ___faces, ___edges, ___vertices.
   Are these two shapes dual
   to each other? (See C13e.)

2. The shapes or "solids" we have
   constructed in the last few
   pages, the tetrahedron, cube,
   octahedron, dodecahedron, and
   icosahedron, are called the
   "regular solids." Because of
   all their symmetries, they
   have a special place in
   mathematics. Also, they have
   another property that they
   share with many less symmetric
   figures—figure out the number
   $$V - E + F$$
   for each of them, where
       V = number of vertices
       E = number of edges
       F = number of faces.

3. Get a balloon and a magic marker.
   Divide the balloon up into
   "faces" any way you want by
   marking edges and vertices on
   the balloon with the magic marker.
   As in exercise #2, calculate the
   number
   $$V - E + F.$$

4. The rotational symmetries of the
   tetrahedron were of two types
   according to the different kinds
   of positions of the axis of
   rotation. (See ex.#2 of C12e.)
   Find three different kinds of
   rotational symmetries of the
   icosahedron using different kinds
   of axes. Make a list of all
   rotational symmetries of the icosahedron using all possible axes of
   rotation. (In 3 dimensions, every
   rotation has an axis of rotation.)

Constructing the baricenter of a triangle

Look at I22 for a moment. The "**baricenter**" we referred to above is the center of the triangle that we found on I22. It is the point where the triangle would balance if you put it on the tip of your finger. Another term we will use on this page is "**median.**" This is any one of the lines through the baricenter of the triangle that we drew on I22. It is any of the three lines in every triangle that passes through a vertex and the midpoint of the side opposite the vertex. To find the baricenter we need only to find the intersection of the medians. However, since all three medians intersect at the baricenter, we only need to find the intersection of two of the medians to find the baricenter. This construction is shown below.

First, take the triangle of which you will find the baricenter.

Find the midpoints of all three sides. Do this by constructing the bisectors of the sides (see C6). The three bisectors are constructed below, separately for clarity.

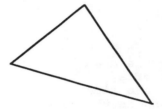

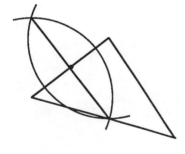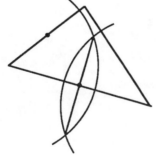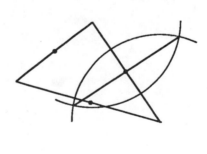

Now take the resulting triangle and connect those three midpoints to the vertices of the triangle opposite the sides that the midpoints are on.

The lines you just drew are the medians of the triangle. Like we said above, you only need to construct two of them to find their intersection, which is the baricenter.

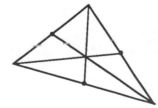

1. In the triangles at the
    right, construct the
    baricenter. It may be
    hard to remember which
    line is which after
    constructing all of those
    bisectors, but it will
    help to keep the radius
    of the arcs you use to
    construct them as small
    as possible.

    Note that no matter what
    shape the triangle is,
    the baricenter is always
    inside the triangle.
    This is because the
    medians always stay inside
    the triangle.

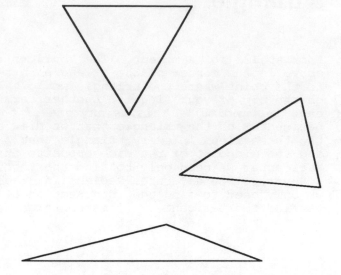

2. Explain why the area
    of the part of a
    triangle on one side
    of the median is the
    same as the area of
    the part on the other
    side of the median.

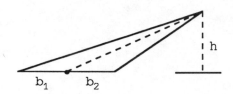

Constructing the altitudes of a triangle

Remember that an **altitude** of a triangle is a line passing through a vertex of the triangle that is perpendicular to the opposite side of the triangle. Notice how different this is from a "median" of a triangle (see C15). Also remember, from I23, that all altitudes of a triangle meet in a common point.

There are a few terms that you must know to understand this page. First of all, an **acute** triangle is a triangle whose angles all measure less than 90˚. An equilateral triangle is acute. An example is at right.

A **right** triangle is a triangle in which one of the angle measures exactly 90˚. See the example at right. Remember that since the angles in a triangle add up to 180˚, no more than one of its angles can be 90˚.

An **obtuse** triangle is one in which one of the angles measures more than 90˚.

To construct the altitudes of a triangle, first take the triangle of which you will construct the altitudes. Now, choose any vertex. Through the vertex, construct a perpendicular to the side opposite this vertex. (See C3e, exercise #1, to construct a perpendicular to a line through a point not on a line). For this example we used an acute triangle, and constructed each altitude separately for clarity.

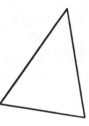

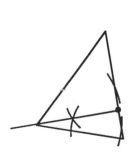

1.   Construct the altitudes
     of the acute triangles
     at right.

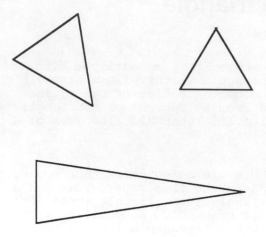

2.   Construct the altitudes
     of the right triangles
     to the right. Notice
     that in a right triangle,
     two of the sides of the
     triangle already "pass
     through a vertex and are
     perpendicular to the side
     opposite the vertex."
     This means that the two
     sides of the triangle
     which form the sides of
     the one 90° angle in the
     triangle are two altitudes
     of the triangle. So, it
     follows that the altitudes
     of a right triangle will
     intersect at the vertex
     where the 90° angle is.

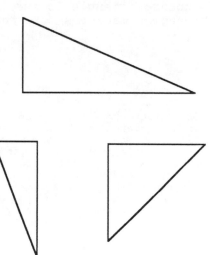

3. Construct the altitudes
of the obtuse triangles
at right. Notice that
in an acute triangle, the
intersection point Z of the
altitudes was well inside
the triangle. In the last
triangle of ex.#1, point
Z got close to the edge
of the triangle as one
of the angles got close
to 90°. In ex.#2, we saw
that when one of the
angles reaches 90°,
point a reaches the edge
of the triangle. So, it
makes sense that, when
one of the angles has
measure greater than
90° (an obtuse triangle),
point Z should go completely
outside the triangle.

Example: The dotted lines
below are the altitudes
of triangle ABC. As you
can see, each altitude passes
through one vertex and
is perpendicular to the
(extension of the)
side opposite that vertex.
You may need to extend
the sides with a straight
edge to make the
construction of a
"perpendicular to a line"
easier.

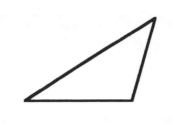

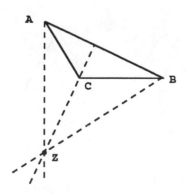

Constructing a circle through three points

One point cannot determine a circle. Given any one point there are many possible circles that could pass through it. See the example, below:

Two points cannot determine a circle either. Given any two points there are many possible circles that could pass through them. See the example, below:

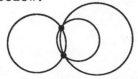

However, three points not on the same line will always determine a unique circle. Notice that in the construction below there is only one intersection of the bisectors, so there can only be one circle through the three points.

To construct the unique circle that passes through three distinct points, first connect the points. Thus, you have three segments, each joining two of the points. Now, construct the perpendicular bisectors of these three segments (see C3e, ex. #3). These three bisectors should intersect in a point (call it point Z). Set your compass to the distance from Z to any one of the three original points (call this distance s) and draw a circle about Z. This is the unique circle that passes through the three points.

Explanation: Since the definition of a circle is that a circle is "the set of points that are at the same distance from a given point," our job in finding the circle passing the three points is to find a point which is equidistant from the three given points. To find points equidistant from the first two of the three given points, we construct the perpendicular bisector of the segment between the points (see C3e, ex.#3). To find points equidistant from the second two of the three given points we construct the perpendicular bisector of the segment between them. The point where these perpendicular bisectors intersect is equidistant from all the three given points.

1. Construct the unique circles
   passing through each set of
   three points at right. For
   clarity each set has been
   separated by a dotted line.

   Notice that the three points
   can be considered as the
   vertices of a triangle. (Just
   join each pair of points
   by a segment.) When we construct a circle through the
   three vertices of a triangle,
   we say that we are <u>circumscribing</u>
   a circle about the triangle.
   There is a unique circle
   circumscribed about any triangle.

2. Explain why you can't
   construct a circle through
   three points if all three
   points lie on a line.

3. Explain why every point P
   on the perpendicular
   bisector of the segment
   AB is equidistant from
   A and B.

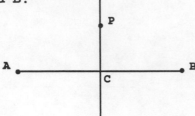

4. Explain why every point Q
   equidistant from A and B
   is on the perpendicular
   bisector.

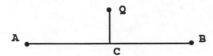

   (Hint: Construct a
   perpendicular to AB
   through Q.)

5. Concurrence theorem for the
   perpendicular bisectors of
   the three sides of a triangle:
   The perpendicular bisectors of
   the sides of any triangle meet
   in a common point.
   Why is this theorem true?
   (Hint: Re-read the explanation
   at the bottom of C17.)

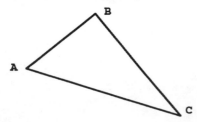

Bisecting a given angle

We want to divide this angle into two
angles, each of which has measure
exactly half the measure of the
given angle:

PT = The point of the compass

PL = The pencil of the compass

To do this, we construct
the **bisector** of the angle.
An angle's bisector is a line drawn
in the exact center of the angle.
The angle made by one side of the
given angle and the bisector is
congruent to the angle made by the
other side of the given angle and
the bisector. To construct the
bisector, set the compass to a
random distance, place the compass
point on the vertex of the given
angle, and mark off the set
distance on each ray of the angle.
Call the marked points on the sides
of the angle A and B. Place the
compass point on A and make an arc on
the interior of the angle, such as
the one at the right. Do the same
with the compass point on B and the
compass setting the same. Connect
the point where the two arcs
intersect to the vertex of the
given angle. This line is the
bisector of the angle.

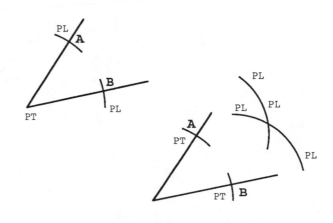

To see why α = β, notice that
the triangles ABD and ACD
are congruent by SSS:

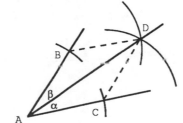

1. Construct the "angle bisector" of the angle at right.

2. Construct the bisectors of each of the sequence of angles at the right:

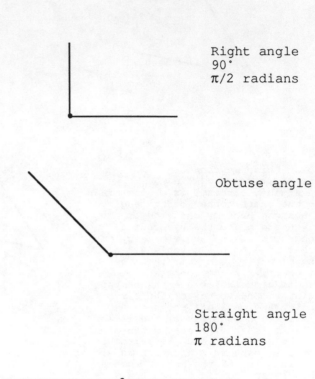

Right angle
90˚
π/2 radians

Obtuse angle

Straight angle
180˚
π radians

Angle of 225˚
or 5π/4 radians.

This
side

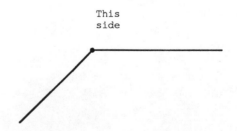

3. The (perpendicular) distance
   from a point to a line
   is the length of a segment
   between the point and the
   line which is perpendicular
   to the line. Explain why
   a point on the bisector of
   an angle is equidistant
   from the two sides of the
   angle.

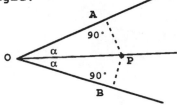

4. Explain why a point which
   is equidistant from the two
   sides of an angle lies on
   the bisector of the angle.

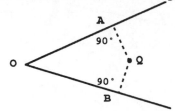

5. Concurrence theorem for
   the bisectors of the angles
   of a triangle:
   The three bisectors of the
   interior angles of a triangle
   meet in a common point.
   Explain why this theorem
   is true.

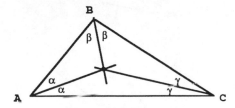

Putting circles inside angles

The object of the following construction is to take a given angle,
and to construct circles that are tangent to each ray of the angle.
In exercise #6 of I25e, we see that
perpendiculars to the rays
at the points of tangency must
pass through the center of any
such circle. So, by exercise
#10 of I18e (HL theorem), $\alpha = \beta$.

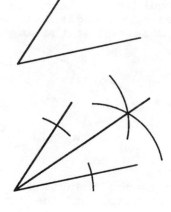

To construct such circles,
we work backwards:

Construct the bisector of the given angle.
(See C18.)

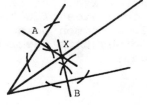

Now choose a point on the bisector.
This is the point we will want
to be the center of our circle.
Call this point "X."   This will be
the center of your circle.
Construct a perpendicular to each
side of the given angle through
this point (see C3e, ex.#1), and
call each point that this
perpendicular intersects the sides
of the angle "A" and "B" respectively.
Set your compass to distance AX
(or BX, it doesn't matter), place the
compass point on X, and draw a circle.
This is the circle we want.

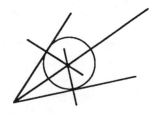

We say that the circle we have constructed is **inscribed** in the
given angle.

1. Inscribe a circle in the angle at right. There are many possible answers, since any point on the angle bisector can be chosen as the center of the circle, and so many different circles could be made.

2. Inscribe a circle in the angles at the right.

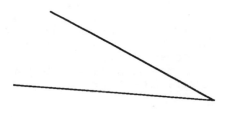

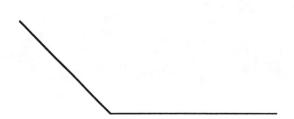

3. Inscribe circles in the
   triangles at the right.
   (Hint: The point in which
   the bisectors of two of
   the angles of the triangle
   intersect will be the center
   of the inscribed circle.
   Bemember that the Concurrence
   Theorem tells us that it
   doesn't matter which two
   angle bisectors we use.)

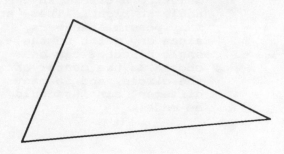

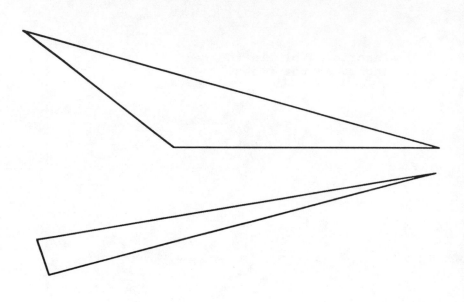

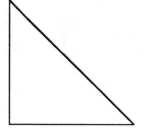

Inscribing circles in polygons

"Inscribing a circle in a polygon" means finding a circle so
that all sides of the polygon are tangent to the circle.
(We will see as we go when it is possible to accomplish this!)
We will try to use what we learned about inscribing circles
in angles in C19, and try to find a circle that is inscribed
perfectly in all angles of the polygon at once. If a circle
is going to be tangent to every side of a polygon,
it will certainly have to be inscribed
in every (interior) angle
of the polygon:

> All our polygons
> are assumed to
> be convex, that
> is, the segment
> between any two
> points inside
> the polygon
> always lies
> entirely inside
> the polygon.

Remember that we can find many circles
inscribed in one of the angles of the
polygon by the construction on C19 (as
long as the angle has measure less than
180°); the center of any such circle
is on the bisector of the angle:

So, if A and B are two angles of
the polygon, the center of the
circle that is inscribed in the
polygon must lie on the dotted
bisector of A, and on the dotted
bisector of B. There is only one
point O that can satisfy both of
these conditions, and that is the
point at the intersection of the two bisectors. Finding the
intersection of only two bisectors is enough—if all the other
angle bisectors don't go through the point O of intersection of the
first two, you won't be able to find a center for a circle
which is inscribed in all the angles of the polygon.

Now that you have found the center O of the inscribed circle, find
the radius by constructing a perpendicular to any side of the polygon
through the point O. By I25e, ex.#6, the point where this perpendicular
intersects the side will be the point on that side where the circle will
be tangent to the side. So, the distance from this point to the
circle must be the radius of the circle. Set your compass to this
distance and draw a circle about the center point O you found. This
circle is the circle that is inscribed in the polygon (if each interior
angles of the polygon has measure less than 180° and all angle
bisectors pass through a common point O).

This example is a 5-gon:
First construct any two
angle bisectors:

Construct perpendicular
to any side through
intersection of bisectors:

Use the distance from X
to Y as the radius and
draw a circle about X:

1. Below is an n-gon with an
    inscribed circle of center
    O.   The vertices of the n-gon
    are A_1, A_2, A_3, ... A_n.
    Explain why the segment OA_1
    is the bisector of angle A_1.
    In other words, explain
    why angle OA_1A_2 is congruent
    to angle OA_1A_n.   (Hint:
    Show that the two right
    triangles in the picture
    are congruent.)

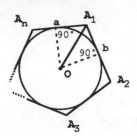

2. Below is an n-gon identical
    to the one in exercise #1.
    Explain why a segment from
    O to any vertex A_* of the
    n-gon is the angle bisector
    of the angle at that vertex.
    (Hint: Show, as you did in
    ex.#1, that angle OA_*A_{*-1} is
    congruent to angle OA_*A_{*+1},
    showing that for any vertex
    A_*, segment OA_* bisects
    angle A_*).

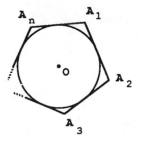

3. Using the information from ex.#2, explain why the conclusion leads to the following "Generalized Concurrence Theorem for Angle Bisectors of a Polygon":
If it is possible to inscribe a circle inside a polygon, then the angle bisectors of the polygon meet in a common point.

4. Looking at the drawing of the n-gon below, assume that the segment OA_1 bisects the angle A_1 of the polygon. Suppose we construct OX_1 and OX_2 such that OX_1 is perpendicular to A_nA_1, and OX_2 is perpendicular to A_1A_2. Explain why $|OX_1| = |OX_2|$.

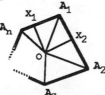

All angles of the polygon have measure < 180°.

5. Looking at the drawing of the n-gon above, assume that the segment OA_2 bisects the angle A_2 of the polygon. Suppose we construct OX_2 and OX_3 such that OX_2 is perpendicular to A_1A_2, and OX_3 is perpendicular to A_2A_3. Explain why $|OX_2| = |OX_3|$.

6. Suppose the n-gon in ex.#4
   has the property that the
   angles bisectors at A_1,
   A_2, ..., A_{n-1} all pass
   through a common point O.
   If OX_1, OX_2, OX_3, ... OX_n
   are perpendiculars to the
   sides of the n-gon passing
   through O, explain why
   $|OX_1| = |OX_2| = ... = |OX_n|$.
   (Hint: Repeat the reasoning
   in exercises #4 and
   #5 over and over.)

7. Use exercise #6 to explain
   the following converse to the
   theorem in ex.#3:   If all
   (except possibly one) of the
   angle bisectors of the polygon
   below meet in one point O,
   one can inscribe a circle in
   the polygon.   (Hint: Name the
   vertices of the polygon A_1,
   A_2, ..., A_n, making A_n the
   vertex whose angle bisector
   you don't know about.)

   All angles
   of the polygon
   have measure
   < 180°.

8. Explain why, if all but one
   of the angle bisectors of the
   above polygon pass through
   the common point O, the last
   angle bisector must pass
   through O also.   (Notice that
   this gives another proof of
   the Concurrence Theorem for
   Angle Bisectors of a triangle—
   see C18e, exercise #5.)

Circumscribing circles about polygons

"Circumscribing a circle about a polygon" means finding a circle so that all vertices of the polygon lie on the circle.
(We will see as we go when it is possible to accomplish this!)
We will try to use what we learned about chords of circles in ex.#2 of C9e, and try to find a circle that
has all of the sides of the given
polygon as chords. If a circle is
going to pass through every vertex
of a polygon, it will certainly have
to have every side of the polygon
as a chord.

Remember that we can find many circles
with a given side of the polygon as
chord; by ex.#3 of C3e and ex.#2
of C9e; the center of any such
circle is on the bisector of the side:

So, if A and B are two sides of
the polygon, the center of the
circle that is circumscribed
about the polygon must lie on the dotted
bisector of A, and on the dotted
bisector of B. (We will assume that
all interior angles of the polygon have
measure less than 180° so that successive
bisectors intersect toward the inside
of the polygon.)   There is only one
point that can satisfy both of
these conditions, and that is the
point O at the intersection of the two bisectors. Finding the
intersection of only two bisectors is enough--if all the other
side bisectors don't go through the point of intersection of the
first two, you won't be able to find a center for a circle
which has each side of the polygon as a chord.

Now that you have found the center O of the circumscribed circle, find
the radius by constructing the distance to any vertex of the polygon
from the point O. We will see in the exercises that all these distances
will be the same. Set your compass to this distance and draw a circle
about the center point O you found. This circle is the circle that is
circumscribed about the polygon (if each interior angle of the polygon
has measure less than 180° and all side bisectors pass
through a common point O).

Our example is a 5-gon.
Construct the perpen-
dicular bisectors of
two sides:

Measure the length
from O to any vertex
(it doesn't matter
which if the con-
struction is possible):

Set your compass to this
length to make a circle about
the center point O:

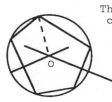

This is the
circle that is
circumscribed
about the
polygon.

1. Below is an n-gon with a
    circumscribed circle with
    center O. The vertices of the
    n-gon are A_1, A_2, A_3, ... A_n.
    Explain why the segment aA_1
    is the bisector of side A_1A_2.
    In other words, explain
    why side aA_1 is congruent
    to side aA_2. (Hint:
    Show that the two right
    triangles in the picture
    are congruent.)

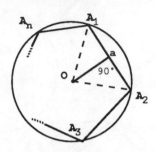

2. Below is an n-gon identical
    to the one in exercise #1.
    Explain why a perpendicular
    from O to any side of the
    n-gon is the bisector
    of that side. (Hint:
    Show, as you did in ex.#1,
    that segment A_xX_x is
    congruent to segment X_xA_{x+1}.)

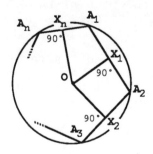

3. Using the information from ex.#2, explain why the conclusion leads to the following "Generalized Concurrence Theorem for Perpendicular Side Bisectors of a Polygon": If it is possible to circumscribe a circle about a polygon, then the side bisectors of the polygon meet in a common point.

4. Looking at the drawing of the n-gon below, assume that the segment OX_1 bisects the side A_1A_2 of the polygon. Explain why $|OA_1| = |OA_2|$.

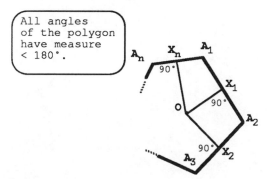

All angles of the polygon have measure $< 180°$.

5. Looking at the drawing of the n-gon above, assume that the segment OX_2 bisects the side A_2A_3 of the polygon. Explain why $|OA_2| = |OA_3|$.

6. Suppose the n-gon in ex. #4
   has the property that the
   perp. side bisectors of A_1A_2,
   A_2A_3, ..., $A_{n-1}A_n$ all pass
   through a common point O.
   If OX_1, OX_2, OX_3, ... OX_n
   are perpendiculars to the
   sides of the n-gon passing
   through O, explain why
   $|OA_1| = |OA_2| = ... = |OA_n|$.
   (Hint: Repeat the reasoning
   in exercises #4 and
   #5 over and over.)

7. Use exercise #6 to explain
   the following converse to the
   theorem in exercise #3: If
   all (except possibly one) of
   the perp. side bisectors of
   the polygon below meet in
   one point O, one can
   circumscribe a circle about
   the polygon. (Hint: Name the
   vertices of the polygon A_1,
   A_2, ..., A_n, making A_nA_1 the
   side whose perpendicular bisector you don't know about.)

All angles
of the polygon
have measure
< 180°.

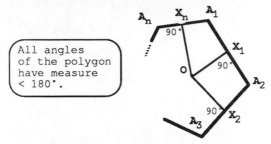

8. Explain why, if all but one
   of the side bisectors of the
   above polygon pass through
   the common point O, the last
   angle bisector must pass
   through O also. (Notice that
   this gives another proof of
   the Concurrence Theorem for
   Side Bisectors of a Triangle—
   see C17e, exercise #5.)

Drawing triangles on the sphere

In order to do this page, you will need a sphere on which to draw.
The easiest way to find a sphere is to take a balloon that is as
close as possible to a sphere. You should use the distance 7cm as
"one unit" in the construction below. So, to get a sphere of radius one
unit, blow up your balloon until its circumference is about 44cm.(=$2\pi r$).
Do this by cutting off a piece of string 44cm long, fitting it loosely
around your half-inflated balloon, and blowing up the balloon until
the string is tight. Seal the balloon.

The objective of the construction below is to construct a
spherical triangle when given that the sides will be 10cm, 6cm,
and 8cm. Basically this is the SSS principle you learned for
triangles in the plane, applied to the sphere. This construction
is not meant to be exact, just to give you the basic idea.

The first step is to construct a
spherical line, or "great circle"
(see I35) to use as a base for your
triangle. Remember that this is
the set of points that a sphere has
in common with a plane that divides
the sphere into two perfect halves.
The easiest way to see this is to
find a globe; the equator of the
earth is a great circle. Draw one
of these on your balloon.

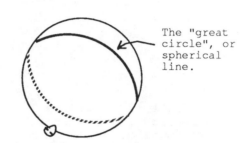

The "great
circle", or
spherical
line.

Now, find any point on the line to
serve as the first vertex of the
triangle. Call this "point A."
Now, cut off a piece of string to
the length of 10cm, the first given
side. Hold one end of the string
on point A, and pull out the other
end, following the line you drew in
the first step, along the curve of
the sphere. When it is pulled tight,
mark "point B" at the other end of
the string.

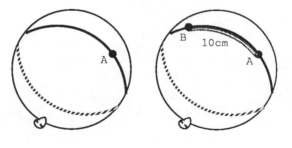

Now, cut off pieces of string to the
lengths of 8cm and 6cm, respectively.
Tape one end of one of them to the point
A and one end of the other to B (it doesn't
matter which one goes with which point).
Pull these pieces of string "tight" over
the curve of the sphere and move them
around until their free ends just meet
(see right). Where they touch at the ends,
mark "point C." (This is the same as
step 5 of C1.) To draw sides AC and BC of
the spherical triangle, hold the strings in place and mark along
them with a pen. Since the strings are pulled tight, they will trace out the
shortest path between their ends—so this path goes along a great circle (see I35).

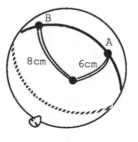
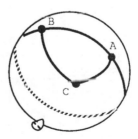

1. Describe the construction
    of a spherical triangle
    when given two angles
    and the side between them.

2. Use a balloon of radius 7
    centimeters to construct
    a spherical triangle with
    sides 8cm, 5cm, and 7cm.

Constructing hyperbolic lines

In I40, we saw that a line in hyperbolic-land is a piece of circle which meets the edge of hyperbolic-land perpendicularly. On this page, we will see how to construct a line through two given points in hyperbolic-land.

Given points P and Q in hyperbolic-land, construct a circle through P and Q which meets the boundary of hyperbolic-land perpendicularly.

1) Construct a point P' on the ray OP such that |OP|·|OP'| = 1. (Construct |OP'| = 1/|OP| by the construction of the quotient of two known lengths given at the bottom of C5).

2) Construct a circle through the points P, Q, and P', according to the instructions in C17. The part of this circle which is inside hyperbolic-land (shown in bold at right) is the unique hyperbolic line through the given points P and Q.

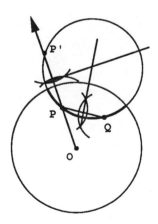

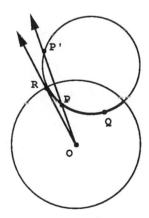

The reason that this construction works is contained in ex.#1, in I25e. Since |OP|·|OP'| = |OR|^2, the ray OR is tangent to the circle we constructed. Since OR is a radius, it is also perpendicular to the boundary of hyperbolic-land. So, the circle we constructed, and the boundary of hyperbolic-land, meet perpendicularly.

1. Construct, with only compass
    and straight edge, the
    hyperbolic line throught the
    points P and Q.

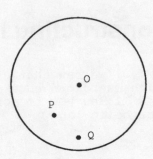

2. Construct, with only compass
    and straight edge, the
    hyperbolic line through the
    points Q and O, the "center"
    of hyperbolic-land. (Hint:
    See I40.)

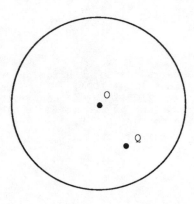

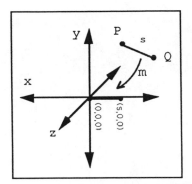

Proof

Distance on the line, motions of the line

Axiom 1:   There is a set called the **(number) line** with elements
called **points**. There is one point on the number line
for each real number, and, in turn, one real number
representing each point on the number line.

Basic example of a number line:

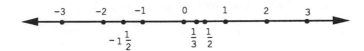

Definition 1:   Given two points   a   and   b   on the number line, the
distance between   a   and   b   is the real number:

$$\sqrt{(a-b)^2} \ = \ |a - b| \ = \ |b - a|$$

Definition 2:   A **motion** of the number line is a rule for assigning
points to other points, so that the original distance
between the points is preserved.

For example, if we call the rule for our motion   m,
m(a) will denote the point on the number line that the
motion assigns to the number   a.   So, we can write the
condition that motions preserve distance as follows:

$$|m(a) - m(b)| \ = \ |a - b|.$$

Example of a motion:

m(x) = x + 1

This motion is the rule that adds one to every number.
For example, to find out where this motion takes the
point 0, put 0 into the formula for the motion, so
m(0) = 0 + 1 = 1. So, the motion moves 0 to 1. Likewise:
m(3) = 3 + 1 = 4, and m(167) = 167 + 1 = 168, and
m(-4) = -4 + 1 = -3, and $m(^1/_2) = \ ^1/_2 + 1 = 1 \ ^1/_2$

As you can see, under
the motion m(x) = x + 1,
all points on the number
line are moved one unit
to the right.

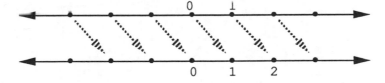

The rule   m   is a motion because it preserves distance:
$$|m(x) - m(y)| \ = \ |(x + 1) - (y + 1)| \ = \ |x + 1 - y - 1| \ = \ |x - y|$$

1. Let m(x) = -x. Where does
   this rule send 1? O? -3?
   $-1\,^1/_2$?

2. For m as in exercise #1,
   what is the distance
   between m(-3) and m(1)?
   Is it the same as the
   distance between -3 and 1?

3. For m as in exercise #1,
   what is the distance
   between m(x) and m(y)?
   Is it the same as the
   distance between x and y?

4. Now, repeat exercises #1,
   #2, and #3, only this time
   use the rule $m(x) = -x + 2$
   as your formula.

5. Explain why $m(x) = 6x$ is
   <u>not</u> a motion.

Distance in the plane

Axiom 2:   There is a set called the **plane** with elements called
points. There is one point in the plane for each pair
of real numbers, and one pair of real numbers for each
point.

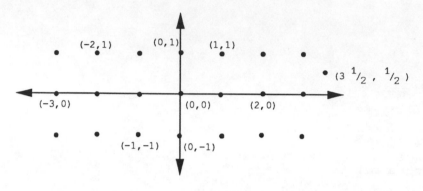

Definition 3:   The **distance** between the points (x_1, y_1) and
(x_2, y_2) in the plane is:

$$\sqrt{(x_2 - x_1)^2 + (y_2 - y_1)^2}$$

> Since $\sqrt{}$ is never negative, distances are never negative numbers.

Notice that, in the formula for distance, it
doesn't matter whether (x_1, y_1) or (x_2, y_2) comes
first:

$$\sqrt{(x_2 - x_1)^2 + (y_2 - y_1)^2} \; = \; \sqrt{(x_1 - x_2)^2 + (y_1 - y_2)^2}$$

since $(x_1 - x_2)^2 = (x_2 - x_1)^2$ and $(y_1 - y_2)^2 = (y_2 - y_1)^2$.

Notation: $d((x_1, y_1), (x_2, y_2))$ will mean "the distance between
(x_1, y_1) and (x_2, y_2)."

Example: $d((3,2), (4,-1)) = \sqrt{(3 - 4)^2 + (2 - (-1))^2} = \sqrt{1^2 + 3^2} = \sqrt{10}$

**1. Find the lengths of the
   sides in the triangles
   shown below:**

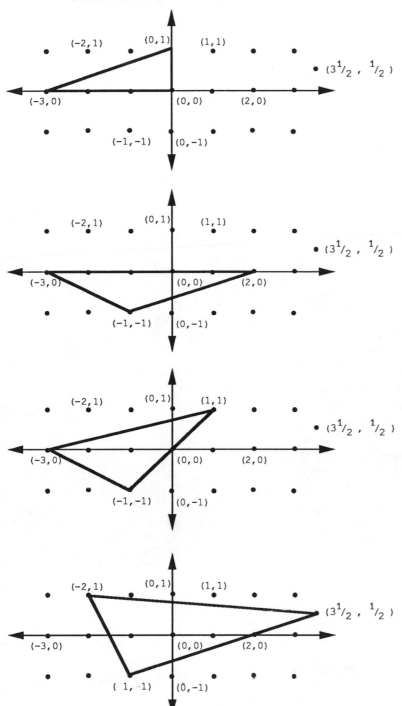

2. The definition of
   distance in the plane is
   motivated by the
   Pythagorean theorem (see
   I9). Use the Pythagorean
   theorem to find the length
   c in the picture below:

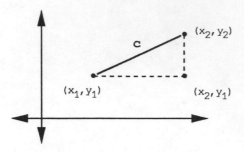

3. The formula for the <u>square</u>
   of the distance between
   two points is easier than
   the formula for the distance
   itself. Show that, if the
   points (x_1,y_1) and (x_2,y_2)
   have coordinates which
   are integers, then the
   square of the distance
   between them is an integer,
   but the distance between
   them need not be an integer.

Motions of the plane

Definition 4:   A **motion** of the plane is a rule for assigning points to other points so that the original distance between the points is preserved.

Example of a motion:

$m(x,y) = (x + 1, -y)$

This rule takes $(2,3)$ to $(2 + 1, -3)$ which is $(3, -3)$. So, $m(2,3) = (3, -3)$.   Also, $m(0,0) = (1,0)$ and $m(-2,-2) = (-1, 2)$. This rule is a motion because:

$$d(m(x_1,y_1),m(x_2,y_2)) = d((x_1 + 1, -y_1),(x_2 + 1, -y_2))$$

$$= \sqrt{((x_2 + 1) - (x_1 + 1))^2 + ((-y_2) - (-y_1))^2}$$

$$= \sqrt{(x_2 + 1 - x_1 - 1)^2 + (-y_2 + y_1)^2}$$

$$= \sqrt{(x_2 - x_1)^2 + (y_2 - y_1)^2}$$

$$= d((x_1,y_1),(x_2,y_2))$$

So, the distance between any point (x_1,y_1) and any other point (x_2,y_2) before they are moved by m is the same as the distance between them after they are moved. So, distances between points are preserved by this rule m. Therefore, this rule ($m(x,y) = (x + 1, -y)$) is a motion.

Example of a motion:

$$m(x,y) = (^1/_2\, x - ^{\sqrt{3}}/_2\, y, \; ^{\sqrt{3}}/_2\, x + ^1/_2\, y)$$

$$m(0,0) = (0,0), \; m(1,0) = (^1/_2, ^{\sqrt{3}}/_2), \; m(0,1) = (-^{\sqrt{3}}/_2, ^1/_2)$$

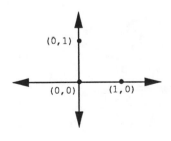

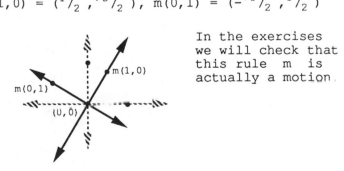

In the exercises we will check that this rule m is actually a motion.

1. Show that the rule
   m(x,y) = (x+5,y-2)
   is a motion of the
   plane.

2. Show that the rule
   n(x,y) = (x-5,y+2)
   is a motion of the
   plane.

3. What happens if you
   do the motion m and
   then follow that
   motion by the motion
   n? (See exercises #1
   and #2.)

4. What happens if you
   do the motion n and
   then follow that
   motion by the motion
   m?

Oldville

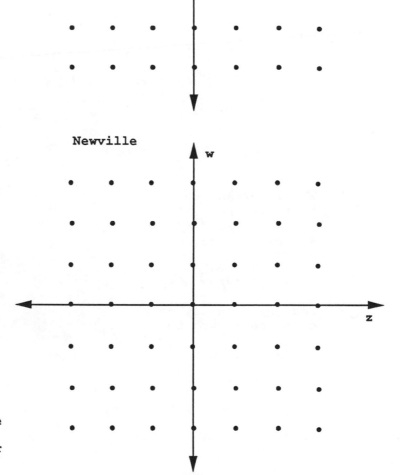

5. **Using your tracing paper copy the (Euclidean) coordinate system at the right (Oldville):**

The motion $m(x,y) =$

$$(\tfrac{1}{2} x - \tfrac{\sqrt{3}}{2} y,$$

$$\tfrac{\sqrt{3}}{2} x + \tfrac{1}{2} y)$$

moves a point (x,y) in Oldville to a point (z,w) in Newville where

$$z = \tfrac{1}{2} x - \tfrac{\sqrt{3}}{2} y$$

and

$$w = \tfrac{\sqrt{3}}{2} x + \tfrac{1}{2} y$$

Newville

Place your tracing of Oldville over Newville in such a way that each point (x,y) is on top of the (z,w) it goes to under the rule m. So $(x,y) = (0,0)$ is on top of $(z,w) = (0,0)$, $(x,y) = (1,0)$ is on top of $(z,w) = (\tfrac{1}{2}, \tfrac{\sqrt{3}}{2})$ and $(x,y) = (0,1)$ is on top of $(z,w) = (-\tfrac{\sqrt{3}}{2}, \tfrac{1}{2})$.

In fact, putting $(0,0)$, $(1,0)$ and $(0,1)$ down in the correct places forces all other (x,y)'s to come down on top of the points given by the rule m. This is because m is a motion and so it doesn't stretch, bend, or rip your tracing paper.

Test this by marking some points (x,y) on your tracing paper, and seeing if the (z,w)'s underneath them have coordinates given by the formula for m.

6. **We are now ready to check that the rule**

$$m(x,y) = (\tfrac{1}{2} x - \tfrac{\sqrt{3}}{2} y, \tfrac{\sqrt{3}}{2} x + \tfrac{1}{2} y)$$

given at the end of P3 is actually a motion. Fill in the missing entries:

> This computation is a bit complicated. The notation will be simpler if we show that the <u>square</u> of the distance is preserved by m. Taking square roots, that's enough to show that distance is preserved.

$$[d(m(x_1,y_1),m(x_2,y_2))]^2 =$$

$$[\ d((\tfrac{1}{2} x_1 - \tfrac{\sqrt{3}}{2} y_1, \tfrac{\sqrt{3}}{2} x_1 + \tfrac{1}{2} y_1), (\tfrac{1}{2} x_2 - \tfrac{\sqrt{3}}{2} y_2, \tfrac{\sqrt{3}}{2} x_2 + \tfrac{1}{2} y_2))\ ]^2$$

$$= \ ((\tfrac{1}{2} x_2 - \tfrac{\sqrt{3}}{2} y_2) - (\tfrac{1}{2} x_1 - \tfrac{\sqrt{3}}{2} y_1)) + ((\ x_2 + \ y_2) - (\ x_1 + \ y_1))^2$$

$$= \ (\tfrac{1}{2}(x_2 - x_1) - \tfrac{\sqrt{3}}{2}(y_2 - y_1))^2 + (\tfrac{\sqrt{3}}{2}(x_2 - x_1) + \tfrac{1}{2}(y_2 - y_1))^2$$

$$= \ (\tfrac{1}{2})^2 (x_2 - x_1)^2 - 2(\tfrac{1}{2})(\tfrac{\sqrt{3}}{2})(x_2 - x_1)(y_2 - y_1) + (\tfrac{\sqrt{3}}{2})^2 (y_2 - y_1)^2$$

$$\qquad + (\tfrac{\sqrt{3}}{2})^2 (x_2 - x_1)^2 + 2(\tfrac{\sqrt{3}}{2})(\tfrac{1}{2})(x_2 - x_1)(y_2 - y_1) + (\tfrac{1}{2})^2 (y_2 - y_1)^2$$

$$= \ ((\tfrac{1}{2})^2 + (\tfrac{\sqrt{3}}{2})^2)(x_2 - x_1)^2 + ((\tfrac{\sqrt{3}}{2})^2 + (\tfrac{1}{2})^2)(y_2 - y_1)^2$$

$$= \ (\qquad)^2 + (\qquad)^2$$

$$= \ [\ d((x_1,y_1),(x_2,y_2))\ ]^2$$

7. **The important thing about the pair of numbers $(\tfrac{1}{2}, \tfrac{\sqrt{3}}{2})$ in exercise #5, was that:**

$$(\tfrac{1}{2})^2 + (\tfrac{\sqrt{3}}{2})^2 = 1.$$

Let (c,s) be any pair of numbers such that:

$$c^2 + s^2 = 1.$$

Show that the rule

$$m(x,y) = (cx - sy, \ sx + cy)$$

is a motion of the plane.

A list of motions of the line

In P1, we saw that the rule m(x) = x+1 is a motion of the number line, that is, it preserves distance between points on the number line. There is nothing magic about the number 1; if we change the rule to m(x) = x+2, the rule will still be a motion—the only difference is that now the rule adds 2 to every point on the number line

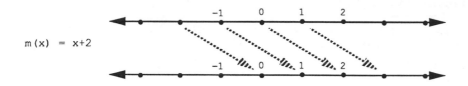

m(x) = x+2

Or we could make the constant we add negative, for example m(x) = x+(-4) = x-4, or m(x) = x - $^9/_2$. The proof that the rule m is a motion is the same, no matter which constant we use. So we let the letter "c" stand for whatever constant we want:

Theorem 1:   For any constant real number c, the rule

$$m(x) = x + c$$

is a motion of the (number) line.

Proof: We must check that this rule preserves the distance between any two points x and y on the line:

$$d(m(x),m(y)) = |\ m(x) - m(y)\ | = |(x+c) - (y+c)|$$
$$= |\ x + c - y - c| = |\ x - y| = d(x,y)$$

where "d(,)" as always means "distance between."

Theorem 2:   The rule m(x) = -x is a motion of the line.

Proof: $$d(m(x),m(y)) = |\ m(x) - m(y)\ | = |(-x) - (-y)|$$
$$= |\ -x + y| = |x - y| = d(x,y)$$

Notice that we can put motions together. Suppose we follow the motion m(x) = -x by the motion n(x) = x+2. We then get the rule n(m(x)) = n(-x) = (-x) + 2. We checked that this was indeed a motion in exercise #4 of P1e.

1. Let m(x) = x-3, and let
    n(x) = x+2. What rule
    do I get altogether if
    I follow the motion m
    by the motion n?

2. Let m(x) = x-3 and n(x)
    = -x. What rule do I
    get altogether if I
    follow m by n?
    What rule do I get
    altogether if I follow
    n by m?

3. Complete the proof of the following:

    Proposition: If a motion of the line m(x) is followed by
                 another motion of the line n(x), the rule

                       p(x) = n(m(x))

                 obtained is again a motion.

    Proof: d(m(x),m(y)) = d(x,y)   since m is a motion.

             d(n(u),n(v)) = d(u,v)   _____

             u and v can be any numbers.
             Suppose u happens to be
             equal to m(x) and v
             happens to be equal to m(y).

             d(n(m(x)),n(m(y))) = d(m(x),m(y))   by substitution.

             d(n(m(x)),n(m(y))) = d(x,y)   by _____.

             So the rule p(x) = n(m(x)) is
             a motion   since, by the previous step,
                                                  it satisfies the condition
                                                  in the definition of a
                                                  motion.

4. Will the Proposition in
    exercise #3 be true if m
    and n are motions of the
    plane instead of motions
    of the line? Explain
    your answer.

A complete list of motions of the line

Theorem 3:   Let m(x) be a motion of the line such that   m(0) = 0.
Then either   m(x) = x   for all   x   (the **identity motion**)
or   m(x) = -x   for all   x.

Proof: d(m(0),m(x)) = d(0,x)   since m is a motion.
d(0,m(x)) = d(0,x)   since m(0) = 0.
|m(x)-0| = |x-0| by the definition of distance.
|m(x)| = |x| by algebra.
m(x) = ± x   by the definition of absolute value.

If m(x) = x for all x on the number line, we have
proved the assertion of the theorem.
If m(x) = -x for all x on the number line, we have
proved the assertion of the theorem.
What we must rule out is the possibility that:
    m(a) = a but not -a for <u>some</u> a on the number line
    m(b) = -b but not b for <u>some</u> <u>other</u> b on the number line.

We will assume now for a moment that this bad possibility
actually happens, and show that that can't be right.

    |m(b) - m(a)| = |b - a| since m is a motion.
    |(-b) - a| = |b - a| since we are assuming that
                      m(b) = -b and m(a) = a.
    |b + a| = |b - a| by algebra.
So, either   b + a = b - a   giving a = -a and so a = 0,
or -(b + a) = b - a   giving -b = b and so b = 0.
But if a = 0, then m(a) = 0 = -a contradicting our assumption,
and if b = 0, then m(b) = 0 = b also contradicting our assumption.

So the bad case can't possibly happen; so the theorem is proved.

Theorem 4:   Let m(x) be any motion of the line. Then m is either one
of the motions listed in Theorem 1, or it is obtained by
following the motion in Theorem 2 by one of the motions
listed in Theorem 1.

Proof: Suppose m(0) = some number c. Define a new motion n(x) = x - c,
and follow the motion m by this new motion n to get the rule
p(x) = n(m(x)). Then p is a motion by exercise #3 in P4e, and
        p(0) = n(m(0)) = n(c) = c - c = 0.
So p satisfies the conditions in Theorem 3. So, by Theorem 3,
either p(x) = x for all x, or p(x) = -x for all x. If p(x) = x,
then n(m(x)) = x, so (m(x)) - c = x, so m(x) = x + c for all x,
and so m is one of the motions listed in Theorem 1.
If p(x) = -x, then (m(x)) - c = -x, so m(x) = -x + c for all x,
and so m is obtained by following the motion in Theorem 2 by
one of the motions listed in Theorem 1.

1. We can make the line
   into the subset of
   all points (x,0) of
   the plane, where x
   can be any real
   number.

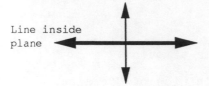

Line inside
plane

   Show that the distance
   between the points x_1
   and x_2 on the line
   is the same as the
   distance between $(x_1,0)$
   and $(x_2,0)$ in the plane.

2. We could also put the
   line inside the plane
   by making it the y-axis.
   Show again that distance
   between points on the
   line is the same whether
   we use the line or plane
   formula for distance.

3. Can you think of other
   ways of putting the
   line inside the plane
   so that line-distance is
   the same as plane-distance?
   If you want to make the
   problem harder, do it so
   that the line is on a slant.

4. In exercise #4 of P1e
    we saw that the rule
    $m(x) = -x + 2$ is a
    motion of the line.
    Theorem 4 says that
    this m is the motion
    of Theorem 2 followed
    by one of the motions
    listed in Theorem 1.
    Which motion from
    Theorem 1 do we use?

5. Suppose $m(x) = -x$, and
    $n(x) = x + 5$. Then
    $p(x) = n(m(n(x)))$ is
    a motion because it is
    made up of motions. So
    by Theorem 4 we must be
    able to write
    $$p(x) = \pm x + c.$$
    What is the sign in front
    of x? What is c?

6. Suppose $p(x) = m(n(m(x)))$
    in exercise #5. Answer
    the questions about p.

Motions of the plane: Translations

Theorem 5:   Let a and b be any two fixed real numbers. The rule
$$m(x,y) = (x+a, \ y+b)$$
is a motion of the plane (called a **translation**).

Proof: $d(m(x_1,y_1), \ m(x_2,y_2)) = d((x_1+a,y_1+b), \ (x_2+a,y_2+b))$

$$= \sqrt{((x_2+a) - (x_1+a))^2 + ((y_2+b) - (y_1+b))^2}$$

$$= \sqrt{(x_2 + a - x_1 - a)^2 + (y_2 + b - y_1 - b)^2}$$

$$= \sqrt{(x_2 - x_1)^2 + (y_2 - y_1)^2}$$

All this
is true
by the
definition
of
distance
in the
plane, and
by algebra.

Definition 5:   The **inverse** of a motion m is a motion n such that,
if we follow m by n, we get back to where we started,
and if we follow n by m, we get back to where we
started. In equation form:

$$n(m(x)) = x \quad \text{for all } x \text{ on the line}$$
or $n(m(x,y)) = (x,y)$ for all (x,y) in the plane,
and

$$m(n(x)) = x \quad \text{for all } x \text{ on the line}$$
or $m(n(x,y)) = (x,y)$ for all (x,y) in the plane.

Theorem 6:   The translation $m(x,y) = (x+a, \ y+b)$ has an inverse
which is also a translation and which is given by the rule:
$$n(x,y) = (x-a, \ y-b)$$
We write n = (inverse m).

Proof: The rule n is a translation by the definition of translation.
(The real number a is replaced by -a and the real number b
is replaced by -b.) Also:

$n(m(x,y)) = n(x+a, \ y+b) = ((x+a) - a, \ (y+b) - b) = (x,y)$ and
$m(n(x,y)) = m(x-a, \ y-b) = ((x-a) + a, \ (y-b) + b) = (x,y)$.

So n satisfies the two conditions in the definition of inverse
motion. So n is the inverse of m.

In the exercises in this section (and later), we will practice composing motions, that is, doing one motion to get from Startville ((x,y)-coordinate land) to Middletown ((z,w)- coordinate land), and then following that by another motion to get from Middletown to Endville ((u,v)- coordinate land).

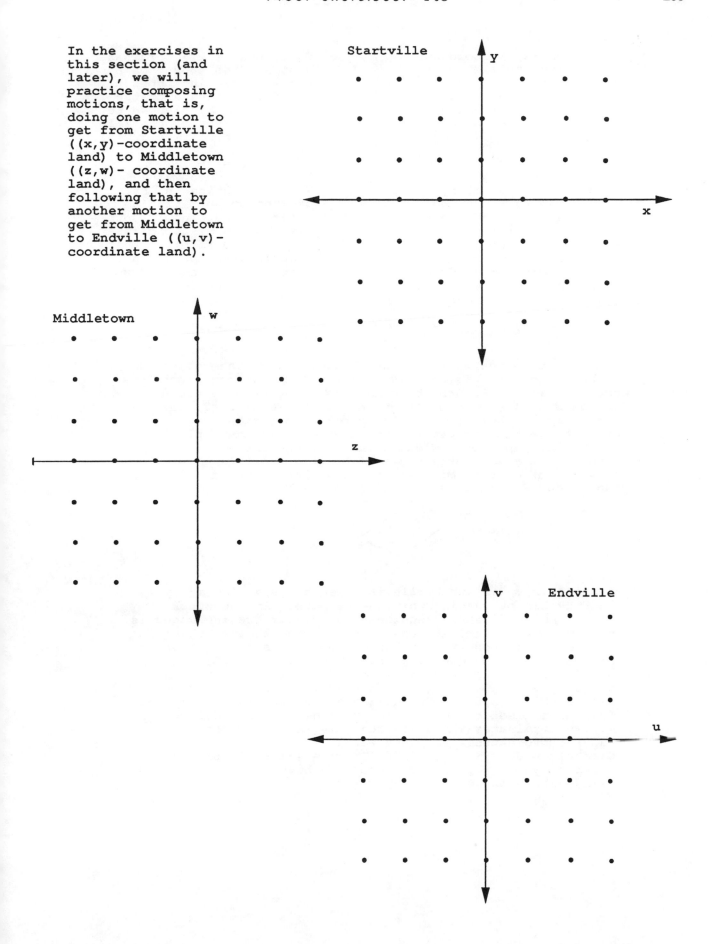

1. Trace the (x,y)-coordinate system (Startville) on a piece of tracing paper. Suppose we have a translation from Startville to Middletown given by the rule (z,w) = m(x,y) = (x+3,y-5), that is,

    z = x+3
    w = y-5.

    Make the motion m by picking up your tracing paper and putting it down over the (z,w)-coordinate system (Middletown) so that each (x,y)-point on the tracing paper comes down on top of the (z,w)-point in Middletown for which

    z = x+3
    w = y-5.

2. Trace the (z,w)-coordinate system (Middletown) on a piece of tracing paper. Suppose we have a translation from Middletown to Endville given by the rule (u,v) = n(z,w) = (z-3,w+5), that is,

    u = z-3
    v = w+5.

    Make the motion n by picking up your tracing paper and putting it down over the (u,v)-coordinate system (Endville) so that each (z,w)-point on the tracing paper comes down on top of the (u,v)-point in Middletown for which

    u = z-3
    v = w+5.

3. Now we want to see what happens when we follow the motion m by the motion n. To do this we will need a strong light under Endville. Put the tracing paper copy of Middletown down on Endville by the motion n as is exercise #2. (Use paper clips to hold the tracing paper in position.) Now, put the tracing paper copy of Startville down on the tracing paper copy Middletown as in exercise #1. So we have a stack three pages high, with Startville ((x,y)-land) on top, Middletown ((z,w)-land) in the middle, and Endville ((u,v)-land) on the bottom:

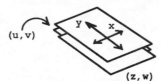

A point (x,y) in Startville is taken to (u,v) = n(m(x,y)) in Endville by passing through the papers in the stack to the point directly underneath. If your tracing paper is transparent enough, and the light underneath everything is strong enough, check that the motion n is the inverse of the motion m by checking that n(m(0,0)) = (0,0), n(m(6,2)) = (6,2), n(m(-4,3)) = (-4,3), in fact, for any point (x,y), (u,v) = n(m(x,y)) = (x,y). That is, if we start with a point (x,y) in Startville and do the motion m, and then do the motion n, the point we get to in Endville has (u,v)-coordinates which are the same as the original (x,y)-coordinates. In other words, (u,v) = (9,7) is directly under (x,y) = (9,7), (u,v) = (-1,8) is directly under (x,y) = (-1,8), etc., etc. We have that, if (u,v) = n(m(x,y)), then u = x and v = y.

4. Verify what we did in exercises #1, 2 and 3 by algebra:

   $z = x+3$ $u = z-3$

   $w = y-5$ $v = w+5$

   Use substitution to give a formula for u and v in terms of x and y.

5. What is the inverse motion of the translation

   $m(x,y) = (x+10, y+9)$?

6. What is the inverse motion of the translation

   $m(x,y) = (x-10, y-9)$?

Motions of the plane: Rotations

Theorem 7:   Let c and s be two real numbers such that
$$c^2 + s^2 = 1.$$
The rule

$$m(x,y) = (cx-sy, \; sx+cy)$$

is a motion of the plane (called a **rotation**).

Proof: What must be checked is that
$$d(m(x_1,y_1),m(x_2,y_2)) = d((x_1,y_1),(x_2,y_2))$$
or, what is the same, that
$$d(m(x_1,y_1),m(x_2,y_2))^2 = d((x_1,y_1),(x_2,y_2))^2$$

But this is exactly what we did in exercises #6 and #7 of P3:.

$$[d(m(x_1,y_1),m(x_2,y_2))]^2 =$$

$$[d((cx_1 - sy_1, \;\; sx_1 + cy_1), \; (cx_2 - sy_2, \;\; sx_2 + cy_2)) \;]^2$$

$$= \;\; ((cx_2 - sy_2)-(cx_1 - sy_1))^2 + ((sx_2 + cy_2)-(sx_1 + cy_1))^2$$

$$= \;\; (c(x_2 - x_1)-s(y_2 - y_1))^2 + (s(x_2 - x_1)+c(y_2 - y_1))^2$$

$$= \;\; c^2(x_2 - x_1)^2 -2\, cs(x_2 - x_1)(y_2 - y_1) + s^2(y_2 - y_1)^2$$
$$+\; s^2(x_2 - x_1)^2 + 2sc(x_2 - x_1)(y_2 - y_1) + c^2(y_2 - y_1)^2$$

$$= \;\; (c^2 + s^2)(x_2 - x_1)^2 + (s^2 + c^2)(y_2 - y_1)^2$$

$$= \;\; (x_2 - x_1)^2 + (y_2 - y_1)^2$$

$$= \;\; [\; d((x_1,y_1),(x_2,y_2)) \;]^2$$

Theorem 8:   The rotation $m(x,y) = (cx-sy, \; sx+cy)$ has an inverse which
is also a rotation and is given by the rule
$$n(x,y) = (cx+sy, \; -sx+cy).$$
We write n = (inverse m).

Proof: $n(m(x,y)) = n(cx-sy, \; sx+cy)$
$$= (c(cx-sy)+s(sx+cy), \; -s(cx-sy)+c(sx+cy))$$
$$= ((c^2+s^2)x, \; (s^2+c^2)y)$$
$$= (x,y)$$

$m(n(x,y)) = m(cx+sy, \; -sx+cy)$
$$= (c(cx+sy)-s(-sx+cy), \; s(cx+sy)+c(-sx+cy))$$
$$= ((c^2+s^2)x, \; (s^2+c^2)y)$$
$$= (x,y)$$

1. Show that every rotation
    of the plane takes $(0,0)$
    to $(0,0)$ and takes $(1,0)$
    to (c,s).

2. Show that $c^2 + s^2 = 1$
    exactly when (c,s) is
    a point on the circle
    of radius 1 around $(0,0)$
    (called the <u>unit circle</u>).

3. Is there a rotation which
    takes $(1,0)$ to $(\sqrt{2}/2, \sqrt{2}/2)$?
    If so, what is its formula?

4. Is there a rotation which
    takes $(\sqrt{2}/2, \sqrt{2}/2)$ to $(1,0)$?
    If so, what is its formula?

5. Where does the rotation
    n in exercise #4 take
    (1,0)?

6. Verify exercises #4 and
    #5 with tracing paper—
    put
    (z,w) = n(x,y)
    = (($\sqrt{2}/2$)x+($\sqrt{2}/2$)y, -($\sqrt{2}/2$)x+($\sqrt{2}/2$)y)
    and make a tracing paper
    copy of the (x,y)-coordinate
    system, and put it on top
    of the (z,w)-coordinate
    system so that (1,0) is
    on top of ($\sqrt{2}/2$, -($\sqrt{2}/2$))
    (and, of course, (0,0)
    is on top of (0,0)).
    See that ($\sqrt{2}/2$, $\sqrt{2}/2$) is
    on top of (1,0).

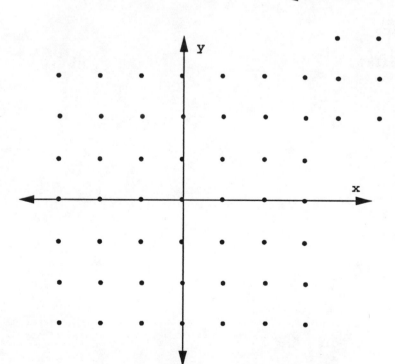

Motions of the plane: Vertical flip

Theorem 9:   The rule $m(x,y) = (x, -y)$ is a motion of the plane.

Proof: $d(m(x_1,y_1),m(x_2,y_2)) = d(x_1,-y_1),(x_2, -y_2))$

$$= \sqrt{(x_2 - x_1)^2 + ((-y_2)-(-y_1))^2}$$

$$= \sqrt{(x_2 - x_1)^2 + (y_2 - y_1)^2}$$

$$= d((x_1,y_1),(x_2,y_2)).$$

Theorem 10:   The vertical flip $m(x,y) = (x, -y)$ has an inverse which is the vertical flip itself. That is,

$$m(m(x,y)) = (x,y).$$

Proof: There is only one thing to check this time:

$$m(m(x,y)) = m(x, -y) = (x, -(-y)) = (x,y).$$

We now have three kinds of motions of the plane, translations, rotations, and the vertical flip. Each of these motions has an inverse motion. In the next two pages we want to show that we can get all motions of the plane by following a rotation by a translation, or by following a vertical flip by a rotation and then following that by a translation.

Once we show this, it will be easy to see that every motion of the plane has an inverse. It's just like with knots—to untie any knot, you do the opposite step for each step you did to tie the knot, and you do them in the opposite order. So, for example,the inverse motion to vertical flip followed by a rotation followed by a translation is the inverse translation followed by the inverse rotation followed by the vertical flip.

1. Given a point (a,b) in
   the plane, find a motion
   of the plane which takes
   (a,b) to (0,0). Find
   a motion of the plane
   which takes (0,0) to
   (a,b).

2. Given a point (e,f) in
   the plane, let
   $d = d((0,0),(e,f))$

   $= \sqrt{e^2 + f^2}$

   Show that (e/d, f/d) is
   on the unit circle,
   that is,
   $$(e/d)^2 + (f/d)^2 = 1.$$

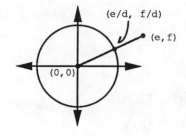

3. Given the points (0,0)
   and (e,f) in the plane,
   find a motion n of
   the plane which leaves
   (0,0) fixed and is such
   that
   $$n(e,f) = (d,0)$$
   for some positive number
   d. (Hint: Use exercise
   #2.)

4. Given two distinct points
   (a,b) and (a',b') in the
   plane, find a motion m
   of the plane so that
   $$m(a,b) = (0,0)$$
   $$m(a',b') = (d,0)$$
   for some positive number d.
   (Hint: Use exercise #1 and
   then exercise #3.)

5. Show that, in exercise #4,
   $d = d((a,b),(a',b'))$.

6. Given the points (a,b) and
   (a',b') in exercise #4,
   find another motion p,
   different from m, so that
   it is again true that
   $$p(a,b) = (0,0)$$
   $$p(a',b') = (d,0).$$

7. Given two distinct points
   (a,b) and (a',b') in the
   plane, find two motions
   of the plane, both of
   which take (0,0) to (a,b)
   and take (d,0) to (a',b').
   (Remember: d has to be
   the distance between (a,b)
   and (a',b') for this to
   be possible.)

Motions of the plane fixing (0,0) and (a,0)

Theorem 11: Let m be a motion of the plane such that
        m(0,0) = (0,0) and m(a,0) = (a,0)
    for some fixed positive number a.
    Then either m is the identity motion
            m(x,y) = (x,y)
    or m is the vertical flip
            m(x,y) = (x, -y).

Proof: Let (x,y) be some point in the plane. m takes it somewhere,
    say m takes it to a point (z,w):
                m(x,y) = (z,w).
    Since m leaves (0,0) fixed, we have:

    d(m(x,y),m(0,0)) = d((x,y),(0,0))   since m is a motion.
    d((z,w),(0,0)) = d((x,y),(0,0))   by substitution.

    $\sqrt{(z-0)^2 + (w-0)^2} = \sqrt{(x-0)^2 + (y-0)^2}$ by definition of dist.

    $z^2 + w^2 = x^2 + y^2$ by algebra.   (1)

    Since m leaves a point (a,0) fixed, we have:

    d(m(x,y),m(a,0)) = d((x,y),(a,0))
    d((z,w),(a,0)) = d((x,y),(a,0))

    $\sqrt{(z-a)^2 + (w-0)^2} = \sqrt{(x-a)^2 + (y-0)^2}$

    $z^2 - 2az + a^2 + w^2 = x^2 - 2ax + a^2 + y^2$ (2)

    The rest of the proof is pure algebra:

    $-2az + a^2 = -2ax + a^2$ subtracting equation (1)
                                        from equation (2).
    $x = z$ by algebra, since a ≠ 0.

    $x^2 + w^2 = x^2 + y^2$ by substitution into
                                        equation (1).
    $w^2 = y^2$ by algebra.

    $w = \pm y$ by algebra.

So, for any point (x,y), either m(x,y) = (z,w) = (x,y) or
m(x,y) = (z,w) = (x, -y). But, to prove the theorem, we must
show something a little stronger, namely, we must show that
either m(x,y) = (x,y) for <u>all</u> points (x,y) in the plane,
or m(x,y) = (x, -y) for <u>all</u> points in the plane.

So, what we must rule out is the possibility that,
$m(x_1,y_1) = (x_1,y_1)$ but not $(x_1, -y_1)$ for some (x_1,y_1)
and $m(x_2,y_2) = (x_2, -y_2)$ but not (x_2,y_2) for some other (x_2,y_2).

We will do this in the first exercise.

1. **Finish the proof of Theorem 11. The model for the reasoning that will be needed is given in the last half of the proof of Theorem 3. Fill in the missing steps in the reasoning given to the right:**

Suppose the bad possibility mentioned at the bottom of P9 actually happens:

$$d(m(x_1,y_1),m(x_2,y_2)) = d((x_1,y_1),(x_2,y_2))$$

So, by substition:

$$d((x_1,y_1),(x_2,-y_2)) = d((x_1,y_1),(x_2,y_2))$$

and, by _____:

$$\sqrt{(x_2-x_1)^2 + ((-y_2)-y_1)^2} =$$
$$\sqrt{(x_2-x_1)^2 + (y_2-y_1)^2}$$

and so, by _____:

$$(-y_2-y_1)^2 = (y_2-y_1)^2.$$

So,

either $y_2+y_1 = y_2-y_1$, so $y_1 = 0$,

or $-y_2-y_1 = y_2-y_1$, so $y_2 = 0$.

But, if $y_1=0$, $m(x_1,y_1) = (x_1,0) = (x_1,-y_1)$ contradicting the assumption at end of P9.
And, if $y_2=0$, $m(x_2,y_2) = (x_2,0) = (x_2,y_2)$ contradicting the assumption at end of P9.

So the bad possibility we were worried about at the end of P9 can't happen.
So Theorem 11 is proved.

2. **Using the coordinates for Oldville and Newville in exercise #5 of P3e and your tracing paper, copy the (x,y)-coordinate system of Oldville onto your tracing paper and mark the points (0,0) and (2,0) on the tracing paper. "Show" Theorem 11 by putting your tracing paper down on the (z,w)-coordinate system of Newville in all possible ways so that (x,y)=(0,0) comes down on top of (z,w)=(0,0) and (x,y)=(2,0) comes down on top of (z,w)=(2,0).**

Remember, you can't stretch, bend or rip your tracing paper.

A complete list of motions of the plane

Theorem 12:   Let m be any motion of the plane. Then m is
either a rotation followed by a translation,
or a vertical flip followed by a rotation followed
by a translation.

Proof: Let m be any motion of the plane whatsoever. Then
$$m(0,0) = \text{some point } (e,f).$$
Let n be the motion given by the formula
$$n(x,y) = (x-e, y-f).$$
Then $n(m(0,0)) = n(e,f) = (e-e, f-f) = (0,0).$

So, if we let p denote the motion we get by doing m first
and then following that by n, $p(0,0) = (0,0).$

Now $p(1,0) = \text{some point } (c,s).$
$d(p(1,0),p(0,0)) = d((1,0),(0,0))$ since p is a motion.

$d((c,s),(0,0)) = d((1,0),(0,0))$ by substitution.

$\sqrt{(c-0)^2 + (s-0)^2} = \sqrt{(1-0)^2 + (0-0)^2}$ by definition of distance.

$c^2 + s^2 = 1$ by algebra.

So, by Theorem 7 and Theorem 8, we can define a rotation
$$q(x,y) = (cx+sy, -sx+cy)$$
Then $q(p(0,0)) = q(0,0) = (0,0)$
and $q(p(1,0)) = q(c,s) = (c^2+s^2, -sc+cs) = (1,0)$
So, if we let k denote the motion we get when we follow the
motion p by the motion q, then
$$k(0,0) = (0,0) \quad \text{and} \quad k(1,0) = (1,0).$$

So Theorem 11 applies to the motion k.   Therefore
either $k(x,y) = (x,y)$ for all (x,y)
or $k(x,y) = (x, -y)$ for all $(x,y).$

But what is the motion k?
$$k(x,y) = q(p(x,y)) = q(n(m(x,y)))$$

So, if $q(n(m(x,y))) = (x,y)$ for all $(x,y),$
then, applying the inverse of q to both sides of this equation,
$(\text{inverse } q)(q(n(m(x,y)))) = (\text{inverse } q)(x,y).$
(By Theorem 8, (inverse q) is a rotation which undoes q.)

So, $n(m(x,y)) = (\text{inverse } q)(x,y).$
Again, $(\text{inverse } n)(n(m(x,y))) = (\text{inverse } n)((\text{inverse } q)(x,y)),$
so, for all $(x,y),$ $m(x,y) = \underbrace{(\text{inverse } n)}_{\text{Translation}}\underbrace{((\text{inverse } q)}_{\text{Rotation}}(x,y)).$

Therefore, the motion m is a rotation followed by a translation.

Similarly, if $q(n(m(x,y))) = (x, -y)$ for all $(x,y),$
then for all $(x,y),$ $m(x,y) = \underbrace{(\text{inverse } n)}_{\text{Translation}}\underbrace{((\text{inverse } q)}_{\text{Rotation}}\underbrace{(x, -y))}_{\text{Flip}}$

So, m is one of the two kinds of motions claimed in Theorem 12.

Give the formula for
the motion of the plane
which takes the old
figure to the new figure:

1.

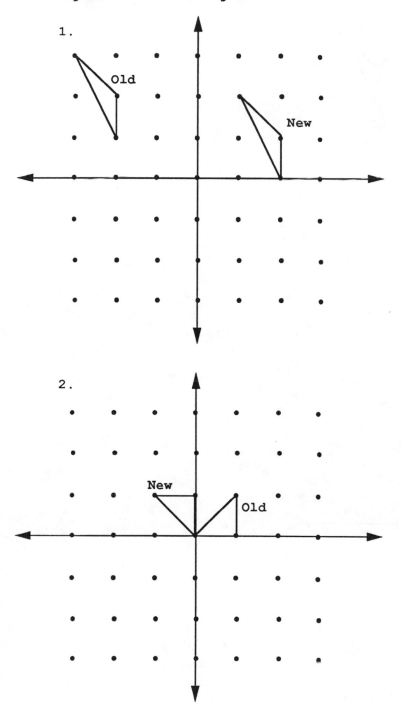

2.

3.

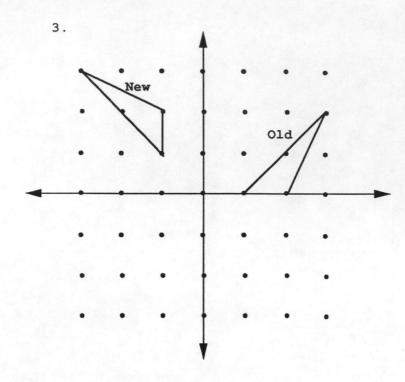

4.

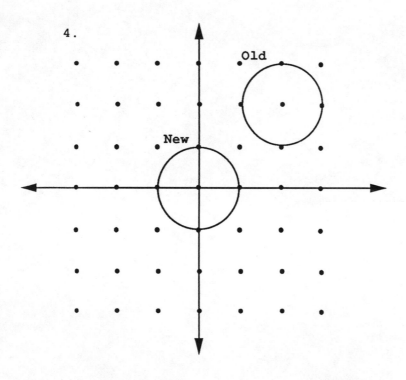

5.

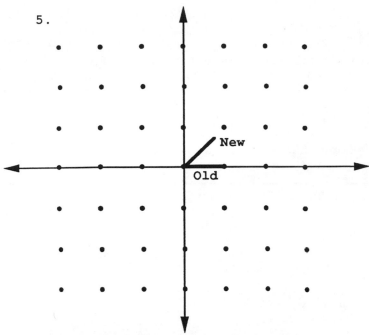

We get the new figure above from
the old by rotating 45° in a
counter-clockwise direction.

6.

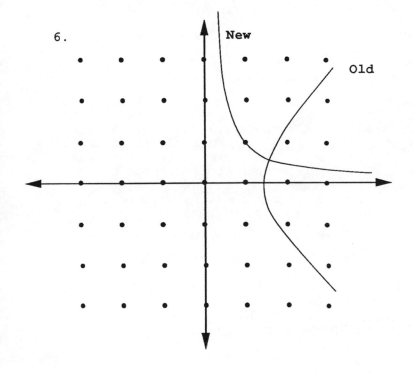

Distance in space

Axiom 3:   There is a set called **(Euclidean) space** with elements
called **points**. There is one point in space for each triple
of real numbers, and one triple of real numbers for each
point in space.

Definition 6:   The **distance** between the points (x_1, y_1, z_1) and (x_2, y_2, z_2)
in space is:

$$\sqrt{(x_2-x_1)^2 + (y_2-y_1)^2 + (z_2-z_1)^2}$$

Theorem 13:   Let A, B, and C, be points (on the line, in the plane,
or in space). Let $d(A,B)$ mean the distance between
A and B. Then:
   1) $d(A,B) = 0$ if and only if $A = B$.
   2) $d(A,B) = d(B,A)$.
   3) $d(A,C) \leq d(A,B) + d(B,C)$.

Proof: We will do the proof for points in space. For points on the
line, or in the plane, the proof goes the same way.
Then $A = (x_1, y_1, z_1)$, $B = (x_2, y_2, z_2)$, and $C = (x_3, y_3, z_3)$.

1) $d(A,B) = 0$ if and only if $\sqrt{(x_2-x_1)^2 + (y_2-y_1)^2 + (z_2-z_1)^2} = 0$

> "P if and only if Q"
> means "P implies Q"
> <u>and</u> "Q implies P"
> (see I20 for example)

if and only if $(x_2-x_1)^2 + (y_2-y_1)^2 + (z_2-z_1)^2 = 0$

if and only if $(x_2-x_1)^2$, $(y_2-y_1)^2$, and $(z_2-z_1)^2$
are all equal to zero.

if and only if $x_1 = x_2$, $y_1 = y_2$, and $z_1 = z_2$, that is $A = B$.

2) $d(A,B) = d(B,A)$ because, in the formula for distance, it does
not matter whether (x_1, y_1, z_1) or (x_2, y_2, z_2) comes first. (See P2.)

3) The proof that $d(A,C) \leq d(A,B) + d(B,C)$ is more complicated.
We'll use what we know about motions to do it. We'll find a
motion m of space in P12e that moves A to $(a,0,0)$, B to $(0,0,0)$
and C to $(c,e,0)$. Then, in P13, we'll show that

$d((a,0,0),(c,e,0)) \leq d((a,0,0),(0,0,0)) + d((0,0,0),(c,e,0))$.

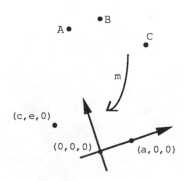

But motions preserve distance, so
$d(A,B) = d((a,0,0),(0,0,0))$,
$d(B,C) = d((0,0,0),(c,e,0))$ and
$d(A,C) = d((a,0,0),(c,e,0))$.

So, by substitution,
$d(A,C) \leq d(A,B) + d(B,C)$.

1. If A = (0,1,5), B = (-1,0,3),
    and C = (2,-1,-1), plot the
    points A, B, and C in the
    three-dimensional coordinate
    system at the right.

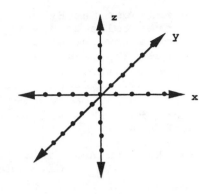

2. Graph the set given by the
    equation z = 2 in the
    coordinate system below:

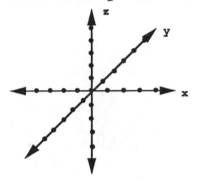

3. In the coordinate system
    below, graph the set
    given by the two equations
        z = 2
        x + y = 0.

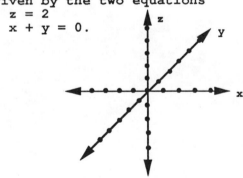

4. Find d(A,B), d(B,C), and
    d(A,C) for the points A,
    B, and C, whose coordinates
    are given in exercise #1.

5. We will use the Pythagorean
   theorem twice to see why the
   formula for the distance d
   between (x_1, y_1, z_1) and
   (x_2, y_2, z_2) should be

$$\sqrt{(x_2 - x_1)^2 + (y_2 - y_1)^2 + (z_2 - z_1)^2}.$$

   (Compare with exercise #2
   of P2e.)

   We start in the horizontal
   plane $z = z_1$. Give the
   formula for the distance c
   between the two points
   (x_1, y_1, z_1) and (x_2, y_2, z_1)
   in this plane:

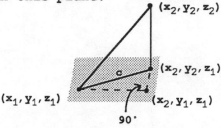

   Next show that the distance
   d we want is given by

$$\sqrt{c^2 + (z_2 - z_1)^2}$$

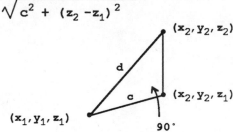

   Finally substitute the formula
   for c from the first step
   to get that d =

$$\sqrt{(x_2 - x_1)^2 + (y_2 - y_1)^2 + (z_2 - z_1)^2}.$$

Motions of space

Definition 7:   A **motion** of space is a rule...   (Fill in just like Definitions 2 and 4).

Theorem 14:   Let a, b, and c be any three fixed real numbers. The rule
$$m(x,y,z) = (x + a,\ y + b,\ z + c)$$
is a motion of space (called a **translation**).

Proof: You fill in the proof. See Theorem 5.

Theorem 15:   Let c and s be two real numbers such that
$$c^2 + s^2 = 1$$
The rule
$$m(x,y,z) = (cx-sy,\ sx+cy,\ z)$$
is a motion of space (called a **rotation**).

Proof: Paste z_1 and z_2 into the proof of Theorem 7:

$$[d(m(x_1,y_1,z_1),m(x_2,y_2,z_2))]^2 =$$

$$[d((cx_1 - sy_1,\ \ sx_1 + cy_1,\ z_1),\ (cx_2 - sy_2,\ \ sx_2 + cy_2,\ z_2))\ ]^2$$

$$= \ ((cx_2 - sy_2)-(cx_1 - sy_1))^2 + ((sx_2 + cy_2)-(sx_1 + cy_1))^2 + (z_2 - z_1)^2$$

$$= \ (c(x_2 - x_1)-s(y_2 - y_1))^2 + (s(x_2 - x_1)+c(y_2 - y_1))^2 + (z_2 - z_1)^2$$

$$= \ c^2(x_2 - x_1)^2 - cs(x_2 - x_1)(y_2 - y_1) + s^2(y_2 - y_1)^2$$
$$+ s^2(x_2 - x_1)^2 + sc(x_2 - x_1)(y_2 - y_1) + c^2(y_2 - y_1)^2 + (z_2 - z_1)^2$$

$$= \ (c^2 + s^2)(x_2 - x_1)^2 + (s^2 + c^2)(y_2 - y_1)^2 + (z_2 - z_1)^2$$

$$= \ (x_2 - x_1)^2 + (y_2 - y_1)^2 + (z_2 - z_1)^2$$

$$= \ [\ d((x_1,y_1,z_1),(x_2,y_2,z_2))\ ]^2$$

Theorem 16:   Let c and s be two real numbers such that
$$c^2 + s^2 = 1.$$
The rule
$$m(x,y,z) = (cx-sz,\ y,\ sx+cz)$$
is a motion of space (called a **rotation**).

Proof: This is just like the proof of Theorem 15. We just rotate the (x,z)-plane leaving the y-axis fixed, instead of rotating the (x,y)-plane leaving the z-axis fixed.

1. There is one more theorem
   which has the same form as
   Theorems 15 and 16. We
   will call it Theorem 16.5.
   Can you state it?

2. Each of the rotations in
   Theorems 15, 16, and 16.5,
   leaves one of the axes of
   space fixed. The axis left
   fixed is called the axis of
   the rotation. What is the
   axis of the rotations in
   Theorem 15? In Theorem 16?
   In Theorem 16.5?

3. Prove the following:
   <u>Proposition</u>: If a motion of
   space $m(x,y,z)$ is followed
   by another motion of space
   $n(x,y,z)$, the rule
   $$p(x,y,z) = n(m(x,y,z))$$
   obtained is again a motion.
   (p is called the <u>composition</u>
   of m with n.)

4. Let $A = (a_1,a_2,a_3)$, $B =$
   (b_1,b_2,b_3), and $C = (c_1,c_2,c_3)$
   be three points of space. Find
   a translation m of space which
   moves A to $(0,0,0)$. Where do
   B and C go under this translation?

5. Now find a rotation n of space
   which moves $m(B)$ to a point
   $(s,0,t)$. (Notice that any rota-
   tion leaves $m(A)=(0,0,0)$ fixed.)
   Hint: Use exercise #3 of P8e and
   Theorem 15.

6. Continuing from exercise
   #5, find a rotation p
   which moves (s,0,t) to a
   point (b,0,0). (Hint: Use
   Theorem 16.)

7. Let r be the motion of
   space which we get when
   we compose the motions of
   exercises #4, 5, and 6,
   that is,
   r(x,y,z) = p(n(m(x,y,z))).
   Find r(A) and r(B).

8. For rotation r in
   exercise #7 and the points
   A, B, and C from exercise
   #4, find a rotation s of
   space that leaves r(A) and
   r(B) fixed and takes r(C)
   to a point (c,e,0). (Hint:
   Use Theorem 16.5.)

9. Prove the following:
   Lemma: Given three points
   A,B, and C in space, there
   is a motion of space which
   moves A to (0,0,0), B to
   a point (b,0,0) with b $\geq$ 0,
   and C to a point (c,e,0)
   with e $\geq$ 0.

10. Use a globe with an (old-
    fashioned) bowl-type holder
    to illustrate the rotations
    of Theorems 15, 16, and 16.5.

The triangle inequality

We are finally ready to finish the proof of the inequality

$$d(A,C) \leq d(A,B) + d(B,C)$$

which we started at the end of P11. This inequality is called the
triangle inequality:

Step 1: Let m be a motion of space such that

  $m(B) = (0,0,0)$, $m(A) = (a,0,0)$ for some $a \geq 0$,
  $m(C) = (c,e,0)$ for some number c and some $e \geq 0$.

  This is possible by exercise #9 of P12e.

Step 2: Show that

  $d((a,0,0),(c,e,0)) \leq d((a,0,0),(0,0,0)) + d((0,0,0),(c,e,0))$

  Proof:

  $-c \leq |c| = \sqrt{c^2} \leq \sqrt{c^2 + e^2}$ by algebra.

  $-2ac \leq 2a\sqrt{c^2 + e^2}$ since $2a \geq 0$.

  $a^2 + c^2 + e^2 - 2ac \leq a^2 + c^2 + e^2 + 2a\sqrt{c^2 + e^2}$

   since adding same thing to both sides preserves $\leq$.

  $(a - c)^2 + e^2 \leq (a + \sqrt{c^2 + e^2})^2$

   by rewriting terms in the previous inequality as
   products.

  $\sqrt{(a - c)^2 + e^2} \leq (a + \sqrt{c^2 + e^2})$

   by taking square roots of both sides.

  But $d((a,0,0),(c,e,0)) = \sqrt{(a - c)^2 + e^2}$

    $d((a,0,0),(0,0,0)) = a$

    $d((0,0,0),(c,e,0)) = \sqrt{c^2 + e^2}$

  So, by substitution,

  $d((a,0,0),(c,e,0)) \leq d((a,0,0),(0,0,0)) + d((0,0,0),(c,e,0))$

  So, substituting from Step 1,

  $d(m(A),m(C)) \leq d(m(A),m(B)) + d(m(B),m(C))$

Step 3: Motions preserve distance, so $d(m(A),m(C)) = d(A,C)$,
    $d(m(A),m(B)) = d(A,B)$ and $d(m(B),m(C)) = d(B,C)$.
    Substituting in the last inequality in Step 2, we get

$$d(A,C) \leq d(A,B) + d(B,C).$$

1. Suppose that A = (1,0,-2),
   B = (3,-1,-1), and C =
   (1,1,1). Find d(A,C),
   d(A,B), and d(B,C). Use
   your calculator to check
   directly that
   d(A,C) ≤ d(A,B) + d(B,C),
   d(A,B) ≤ d(A,C) + d(C,B),
   d(B,C) ≤ d(B,A) + d(A,C).

2. Suppose that A = (1,1,2),
   B = (3,3,6), and C =
   (4,4,8). Graph these three
   points in the coordinate
   system below:

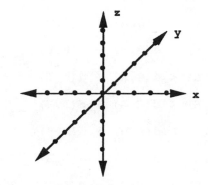

   Show that
   d(A,B) < d(A,C) + d(C,B)
   d(B,C) < d(B,A) + d(A,C)
   but
   d(A,C) = d(A,B) + d(B,C).

3. Can you explain (without
   worrying about proving it)
   why
   d(A,C) = d(A,B) + d(B,C)
   in exercise #2?

4. Let A, B, and C be
   three points of space.

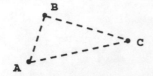

If A, B, and C, make
up the vertices of an
actual triangle, check
off which of the
following are true:

d(A,C) < d(A,B) + d(B,C) __

d(A,B) = d(A,C) + d(C,B) __

d(C,B) ≤ d(C,A) + d(A,B) __

d(C,A) = d(A,B) + d(B,C) __

d(A,B) < d(C,A) + d(C,B) __

d(C,B) < d(C,A) + d(A,B) __

d(A,C) ≤ d(A,B) + d(C,B) __

d(B,A) ≤ d(A,C) + d(C,B) __

d(B,C) = d(B,A) + d(A,C) __.

If A happens to lie on the
segment between B and C,
check which of the following
are true:

d(A,C) < d(A,B) + d(B,C) __

d(A,B) = d(A,C) + d(C,B) __

d(C,B) ≤ d(C,A) + d(A,B) __

d(C,A) = d(A,B) + d(B,C) __

d(A,B) < d(C,A) + d(C,B) __

d(C,B) < d(C,A) + d(A,B) __

d(A,C) ≤ d(A,B) + d(C,B) __

d(B,A) ≤ d(A,C) + d(C,B) __

d(B,C) = d(B,A) + d(A,C) __.

If it happens that B = C,
check which of the statements
at the right are true:

d(A,C) < d(A,B) + d(B,C) __

d(A,B) = d(A,C) + d(C,B) __

d(C,B) ≤ d(C,A) + d(A,B) __

d(C,A) = d(A,B) + d(B,C) __

d(A,B) < d(C,A) + d(C,B) __

d(C,B) < d(C,A) + d(A,B) __

d(A,C) ≤ d(A,B) + d(C,B) __

d(B,A) ≤ d(A,C) + d(C,B) __

d(B,C) = d(B,A) + d(A,C) __.

Co-ordinate geometry is about shapes and more shapes

Line   A line is a subset of space which can be obtained from the x-axis by a motion of (Euclidean) space.

>   Notation: We will sometimes denote the x-axis by
> $\{(x,0,0): x$ real number$\}$ which reads "the set of all
> $(x,0,0)$ such that x is a real number."

Ray   A ray is a subset of space which can be obtained from the positive x-axis by a motion of space. The point corresponding to $(0,0,0)$ under the motion is called the **endpoint** of the ray.

>   Notation: The positive x-axis is $\{(x,0,0): x \geq 0\}$.

Segment   A segment is a set which can be obtained from
$$\{(x,0,0): 0 \leq x \leq d\}$$
by a motion of space. "d" is some (fixed) real number called the **length** of the segment.

Angle   An angle is a set which can be obtained from the following union of two rays
$$\{(x,0,0): x \geq 0\} \text{ and } \{(cx,sx,0): x \geq 0\}$$
by a motion of space. Here we require that
$$c^2 + s^2 = 1.$$
"c" is called the **cosine** of the angle and "s" is called the **sine** of the angle.

Triangle   Same definition as on I2.

Plane   A plane is a set which can be obtained from
$$\{(x,y,0): x \text{ and } y \text{ are real numbers}\}$$
by a motion of space.

Circle   A circle is a set which can be obtained from

$$\{(x,y,o): x^2 + y^2 = r^2\}$$

by a motion of space. "r" is some fixed number called the **radius** of the circle, and the point corresponding to $(0,0,0)$ under the motion is called the **center** of the circle.

Sphere   A sphere is a set which can be obtained from
$$\{(x,y,z): x^2 + y^2 + z^2 = r^2\}$$
by a motion of space. "r" is some fixed number called the **radius** of the sphere, and the point corresponding to $(0,0,0)$ under the motion is called the **center** of the sphere.

1. The definition of angle
   in P14 has a built in
   notion of "direction
   of the angle." We call
   the ray corresponding to
   $\{(x,0,0): x \geq 0\}$ under the
   motion the <u>initial side</u> of
   the angle, and the ray
   corresponding to
   $\{(cx,sx,0): x \geq 0\}$ under the
   motion the <u>final side</u>.
   Draw two angles in the (x,y)-
   plane with initial side
           $\{((x,x): x \geq 1\}$
   having $c = s = \sqrt{2}/2$.

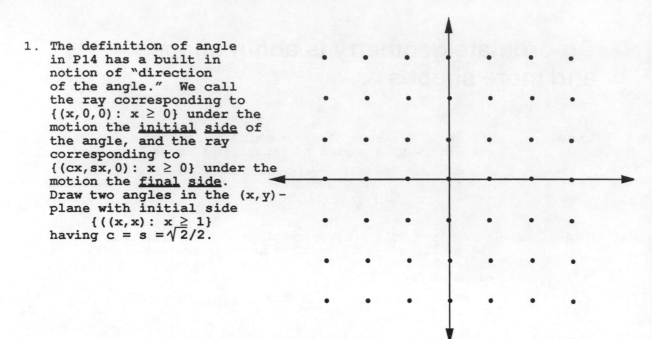

2. There are two possible
   values for the sine of
   this angle:

   **What are
   they?**

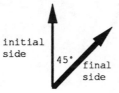

3. When our angles are in a fixed plane, we
   don't want two angles with the same initial
   side and the same cosine and sine. Also we
   want each angle to have a unique number for
   its cosine and a unique number for its sine.
   To achieve this, we require that the angle
   be obtained from
       $\{(x,0): x \geq 0\}$ and $\{(cx,sx): x \geq 0\}$
   by a motion of the plane which is made up
   of rotations and translations only (no vertical
   flips). In the plane below, draw the following
   list of angles:

Initial side	c=	s=
a. $\{(x,x): x \geq 2\}$	0	1
b. $\{(x,x): x \geq -1\}$	$\sqrt{3}/2$	1/2
c. $\{(0,y): y \leq -1\}$	$\sqrt{2}/2$	$\sqrt{2}/2$
d. $\{(0,y): y \leq -1\}$	$\sqrt{2}/2$	$-\sqrt{2}/2$
e. $\{(2x,x): x \geq 1\}$	-1	0

   (Hint: Use I27e or I28e
   or INV COS and INV SIN
   functions on your
   calculator to help find
   angles whose sines and
   cosines you know.)

4. Let $P = (0,0,0)$ and $Q = (s,0,0)$. Suppose R is any point
   (x,y,z) in space. Complete the proof below that
   $$d(P,Q) = d(P,R) + d(R,Q)$$
   only when $R = (x,0,0)$ for some x with $0 \leq x \leq s$.

(1)	$d(P,Q) = d(P,R) + d(R,Q)$	Given
(2)	_____	Definition of distance
(3)	$s \geq \sqrt{x^2} + \sqrt{(s-x)^2} = |x| + |s-x| \geq x + (s-x) = s$	Algebra
(4)	$s = \sqrt{x^2} + \sqrt{(s-x)^2}$	Inequalities in (3) must be equalities
(5)	$\sqrt{x^2+y^2+z^2} + \sqrt{(s-x)^2+y^2+z^2} = \sqrt{x^2} + \sqrt{(s-x)^2}$	Steps (2) and (4)
(6)	$y = z = 0$	------------------
(7)	$0 \leq x \leq s$	$|x| + |s-x| = x + (s-x)$ by Step (3)

5. Define what we should
   mean by the "endpoints"
   of a segment.

6. Given two points P and Q,
   explain why there is a
   segment with endpoints P
   and Q. (Hint: Use exercise
   #9 of P12e.)

7. Suppose a segment with
   endpoints P and Q is given
   by a motion m   of
       $\{(x,0,0): 0 \leq x \leq s\}$.
   Explain why $s = d(P,Q)$.
   (Hint: By definition,
   motions preserve distances.)

8. Suppose R is on a segment
   with endpoints P and Q.
   Explain why
   $d(P,Q) = d(P,R) + d(R,Q)$.
   (Hint: Move the segment
   back to the x-axis.)

The shortest path between two points...

Theorem 17 : **Given two points P and Q in space, suppose there is a point R such that**
$$d(P,Q) = d(P,R) + d(R,Q).$$
Then R lies on every segment with endpoints P and Q.

Proof:

(1) Suppose R is a point such that
d(P,Q) = d(P,R) + d(R,Q), and S is a
segment with endpoints P and Q.

Given.

(2) There is a motion m of space which
takes (0,0,0) to P, (s,0,0) to Q,
and {(x,0,0): 0 ≤ x ≤ s} to S.

Definition of segment.

(3) Some point (a,b,c) goes to R under
the motion m.

(a,b,c) = n(R), where n is the
inverse motion of m. (Here we are
cheating—we never proved that in
space every motion has an inverse!
But this assertion is true. It
follows from the fact that all
motions of space are made up by
combining translations, rotations,
and flips, each of which has an
inverse.)

(4) d((0,0,0),(s,0,0)) = d(P,Q)
d((0,0,0),(a,b,c)) = d(P,R)
d((a,b,c),(s,0,0)) = d(R,Q)

The motion m preserves distances.

(5) d((0,0,0),(s,0,0)) =
d((0,0,0),(a,b,c)) + d((a,b,c),(s,0,0))

Use (4) and the identity
d(P,Q) = d(P,R) + d(R,Q) from (1).

(6) b = 0, c = 0, and 0 ≤ a ≤ x.

Exercise #4 of P14e.

(7) (a,b,c) is in {(x,0,0): 0 ≤ x ≤ s}.

(6)

(8) R is on the segment obtained from
{(x,0,0): 0 ≤ x ≤ s} by the motion m.

R = m((a,b,c)) and (7).

Theorem 18: **Given points P and Q in space, any segment with endpoints P and Q must be the set of all points R such that**
$$d(P,Q) = d(P,R) + d(R,Q).$$
So there is exactly one segment with endpoints P and Q.

Proof:

(1) There is a segment with endpoints P and Q.
Let S denote any such segment.

Ex.#6 of P14e.

(2) If d(P,Q) = d(P,R)+d(R,Q), R is on segment S.

Above theorem.

(3) If R is on segment S, d(P,Q) = d(P,R)+d(R,Q).

Ex.#8 of P14e.

1. Let P and Q be points in
   space. Show that there is
   a path from P to Q of length
   s = d(P,Q).

2. Suppose we have a path from
   P to Q made up of two or more
   segments:

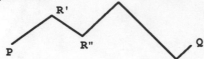

   Explain why there is a
   shorter path made up of
   one fewer segments.

3. Explain why the shortest
   path (of segments) between
   points P and Q is the
   segment with endpoints P
   and Q.

4. Explain why any motion m
   of the plane has a formula
   m(x,y) = (cx-sy+a, sx+cy+b)
   or
   m(x,y) = (cx+sy+a, -sx+cy+b)
   for some constants c, s, a,
   and b. (Hint: Use Theorem 12
   in P10.)

5. A line in the plane is a set
   which can be taken to the
   x-axis by a motion m of the
   plane. Explain why any line
   is given by an equation
       ex + cy + b = 0
   for some constants e, c, and
   b. (Hint: Use exercise #4.)

6. Suppose a set in the plane is given by the equation

   $3x + 4y + 2 = 0$.

   Show that this set is a line. (Hint: Divide both sides of the equation by some number D so that

   $(3/D)^2 + (4/D)^2 = 1$.

   Then use a motion m with

   $s = 3/D$ and $c = 4/D$ and $b = 2/D$ to take the set to the x-axis.)

7. Suppose a set in the plane is given by the equation

   $Ax + By + C = 0$.

   Show that this set is a line. (Hint: Divide both sides of the equation by some number D so that

   $(A/D)^2 + (B/D)^2 = 1$.

   Then use a motion m with

   $s = A/D$ and $c = B/D$ and $b = C/D$ to take the set to the x-axis.)

8. Suppose three points P, Q, and R, in space lie on a line. Show that one of the points must lie on the segment whose endpoints are the other two given points. (Hint: Remember that the line is obtained from the x-axis by a motion. Look at the corresponding points on the x-axis.)

The unique line through
two given points

Theorem 19:   Given two distinct points in space, there is
                   a line passing through the points.

Proof: Use ex.#9 of P12e to find a motion m which
              moves the two points to (0,0,0) and (s,0,0)
              where s = d(P,Q). The inverse motion to m
              takes the x-axis to a line containing the two
              given points.

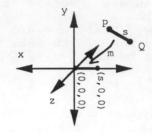

Theorem 20:   Given two distinct points P and Q in space,
                   there is only one line that contains both
                   P and Q.

Proof: Suppose that the theorem is false; we will show that this leads to a
            contradiction, thereby proving the theorem to be true. Suppose there
            are two distinct lines L and L' passing through the two points P and Q.
            Use ex.#9 of P12e to find a motion m which moves P to (0,0,0) and
            Q to (s,0,0). The motion m moves L and L' to two <u>distinct</u> lines
            containing (0,0,0) and (s,0,0). We will be done if we can show that
            the x-axis is the only line containing (0,0,0) and (s,0,0). (Remember
            that since P ≠ Q, then s ≠ 0).

            Suppose K is any line containing (0,0,0) and (s,0,0). Let (a,b,c) be
            a point on K. By ex.#8 of P15e, one of the three points (a,b,c),
            (0,0,0), or (s,0,0) must lie on the segment which has the other two
            as endpoints. For example, suppose (0,0,0) lies on the segment from
            (a,b,c) to (s,0,0). By the second theorem on P15:

   $$d((a,b,c),(s,0,0)) = d((a,b,c),(0,0,0)) + d((0,0,0),(s,0,0))$$

So, by the definition of distance:

$$\sqrt{(a-s)^2 + b^2 + c^2} = \sqrt{a^2 + b^2 + c^2} + \sqrt{s^2}$$
$$(a-s)^2 + b^2 + c^2 = a^2 + b^2 + c^2 + 2s\sqrt{a^2 + b^2 + c^2} + s^2$$
$$a^2 - 2as + s^2 + b^2 + c^2 = a^2 + b^2 + c^2 + s^2 + 2s\sqrt{a^2 + b^2 + c^2}$$
$$-2as = 2s\sqrt{a^2 + b^2 + c^2}$$
$$-a = \sqrt{a^2 + b^2 + c^2}$$
$$a^2 = a^2 + b^2 + c^2$$
$$0 = b^2 + c^2 \qquad \text{that is, both b and c must be equal to zero.}$$

The other two cases are done similarly. So, (a,b,c) is on the x-axis, since
its y-axis and z-axis positions are both 0. Thus, every point on K is on the
x-axis. Since K contains points at all distances from (0,0,0), in fact K must
fill up the entire x-axis. So, we have shown that any line K containing
(0,0,0) and (s,0,0) is the x-axis. This means that m(L) and m(L') must both
be the x-axis, so that L = L' after all. Thus we have the contradiction we
wanted, showing that L and L' are distinct and the theorem is true.

1. In P16 we said that there are three possibilities for positions of the points if $(0,0,0)$, $(s,0,0)$, and (a,b,c) are all on K. We considered the possibility that $(0,0,0)$ was between $(s,0,0)$ and (a,b,c), and we showed that both b and c were zero. We also said that the method for showing $b = c = 0$ in the other two cases was similar. Show that both b and c are zero in the case that $(s,0,0)$ is between $(0,0,0)$ and (a,b,c).

2. Now show that both b and c are zero in the case that (a,b,c) is between $(0,0,0)$ and $(s,0,0)$.

Proving SSS

Using the theorems we have developed in these "P" pages, one can prove all the results about plane and space geometry which we saw in the "I": and "C" pages. We will content ourselves with giving just one example.

Definition: **Two subsets of the plane are congruent if there is a motion of the plane which takes one of the sets exactly to the other.**

Theorem 21:   **If two triangles in the plane have corresponding sides of equal length, then they are congruent.**

Proof:

Given triangles ABC and DEF in the plane with
$$d(A,B) = d(D,E)$$
$$d(B,C) = d(E,F)$$
$$d(C,A) = d(F,D).$$

By exercise #4 of P8e, there is a motion m of the plane such that $m(A) = (0,0)$ and $m(B) = (d,0)$, where $d = d(A,B)$. Also, there is a motion n of the plane such that $n(D) = (0,0)$ and $n(E) = (d,0)$, where $d = d(D,E) = d(A,B)$.

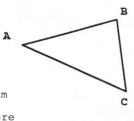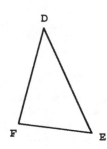

Let $m(C) = (x,y)$ and $n(F) = (x',y')$. Since m and n preserve distances,
$d((0,0),(x,y)) = d(A,C) = d(D,F) = d((0,0),(x',y'))$. Therefore,

$$\sqrt{x^2 + y^2} = \sqrt{x'^2 + y'^2}$$

$$x^2 + y^2 = x'^2 + y'^2 . \qquad (1)$$

In the same way, $d((d,0),(x,y)) = d((d,0),(x',y'))$; so

$$(x-d)^2 + y^2 = (x'-d)^2 + y'^2.$$

$$x^2 - 2xd + d^2 + y^2 = x'^2 - 2x'd + d^2 + y'^2. \qquad (2)$$

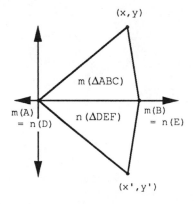

Subtracting (1) from (2) and cancelling like terms, we conclude that $x = x'$. Substituting in (1) and cancelling again, we get $y = y'$, or $y' = \pm y$.

If $(x',y') = (x,y)$, then $m(C) = n(F)$ and so $m(\triangle ABC) = n(\triangle DEF)$, and so $\triangle ABC$ is congruent to $\triangle DEF$.

If $(x',y') = (x,-y)$, follow the motion n by a vertical flip (P8) to move $\triangle DEF$ on top of $m(\triangle ABC)$. Again, in this case, the triangles are congruent.

1. We can do some nice things
   with areas using the ideas
   from the "P" pages, too. To
   begin, given four numbers a,
   b, c, and d, write

$$\begin{vmatrix} a & b \\ c & d \end{vmatrix}$$

   to denote the number
   $$a \cdot d - b \cdot c.$$
   Show that

$$\begin{vmatrix} a & 0 \\ c & d \end{vmatrix}$$

   is the area of the
   parallelogram:

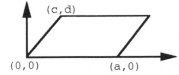

2. Now show that the formula
   for the area of the
   triangle

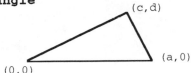

   is:

$$\frac{1}{2} \cdot \begin{vmatrix} a & 0 \\ c & d \end{vmatrix} =$$

$$\frac{1}{2} \cdot \left(\begin{vmatrix} a & 0 \\ c & d \end{vmatrix} + \begin{vmatrix} c & d \\ 0 & 0 \end{vmatrix} + \begin{vmatrix} 0 & 0 \\ a & 0 \end{vmatrix} \right)$$

3. Show that, if m(x,y) is a
   motion of the plane, and if
   (a',b') = m(a,b), etc., then

$$\frac{1}{2} \cdot \left(\begin{vmatrix} a & b \\ c & d \end{vmatrix} + \begin{vmatrix} c & d \\ e & f \end{vmatrix} + \begin{vmatrix} e & f \\ a & b \end{vmatrix} \right) =$$

$$\pm \ \frac{1}{2} \cdot \left(\begin{vmatrix} a' & b' \\ c' & d' \end{vmatrix} + \begin{vmatrix} c' & d' \\ e' & f' \end{vmatrix} + \begin{vmatrix} e' & f' \\ a' & b' \end{vmatrix} \right)$$

   (Hint: Show the formula for a
   translation, for a rotation,
   and for a vertical flip (this
   is where the "-" comes in),
   and then use that every
   motion is a sequence of
   motions of these
   three types.)

4. Use exercises #2 and #3 to
    show that the formula for
    the area of any triangle

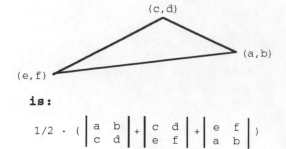

    is:

$$\frac{1}{2} \cdot \left(\begin{vmatrix} a & b \\ c & d \end{vmatrix} + \begin{vmatrix} c & d \\ e & f \end{vmatrix} + \begin{vmatrix} e & f \\ a & b \end{vmatrix} \right)$$

5. Use exercise #4 to give a
    formula for the area of
    an n-gon in terms of the
    coordinates of its vertices.

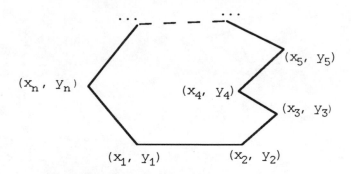

    (Hint: Break the region up
    into triangles and do some
    algebra. To get a "nice"
    formula, you will need to
    use that:

$$\begin{vmatrix} a & b \\ c & d \end{vmatrix} = - \begin{vmatrix} c & d \\ a & b \end{vmatrix}$$

    The formula you get is the
    precursor of a famous
    theorem in two-variable
    calculus called Green's
    Theorem.)

6. We have seen in exercise #4
    that the area of the
    triangle

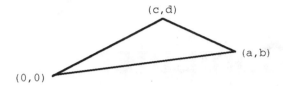

    is

    $$1/2 \cdot \begin{vmatrix} a & b \\ c & d \end{vmatrix} \quad .$$

    Use this formula to
    compute the area of the
    triangle below:

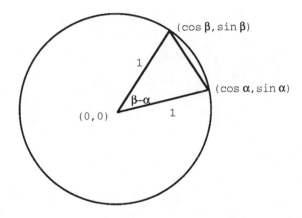

7. Show that the area of the
    triangle in exercise #6 is
    also given by
    $$1/2 \cdot \sin(\beta-\alpha) \quad .$$
    Conclude that
    $\sin(\beta-\alpha) = \sin\beta \cdot \cos\alpha - \sin\alpha \cdot \cos\beta$.
    Compare this formula with
    exercise #10 of I30e.

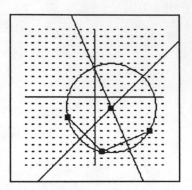

Computer
Programs

Information you'll need
about the CP-pages

This text contains background information on the LOGO language, and information you may need regarding the use of the programs in this book with different versions of LOGO.

Some LOGO background

The LOGO programming language was developed at MIT several years ago in a project headed by Seymour Papert. It was designed to make computer programming as easy as possible, especially for beginners. For this reason, it is used as an introduction to computer programming for children all over the country. In the pages that follow, the language has been taken a step beyond the beginner's viewpoint; it is used to help illustrate many of the ideas and constructions from the book's text that may help students understand the material more fully. For students with some previous experince with LOGO, the programs are intended to be typed into the computer and used according to the directions. For students quite familiar with LOGO, the programs are intended as a challenge: Can you improve on these programs or write other ones to illustrate other geometric concepts? LOGO experts should be able to improve on many of these programs. The CP-pages, as they stand, certainly do not constitute a LOGO textbook, but rather a quick method to see some of the basic concepts of this book illustrated on the computer screen. The LOGO language was chosen because of its superior ability to draw various geometric figures, and manipulate lists of numbers.

The main advantage of LOGO over some other languages is its unique system of drawing on the computer screen, called "turtle graphics." In this system, the user is given control of an imaginary turtle. The user can give commands such as FORWARD, BACK, LEFT, and RIGHT to manipulate the turtle in various ways. The turtle posseses a "pen," which can be placed in the "down" or "up" position. When the pen is down, the turtle leaves a tracing of its path on the screen as it moves, and when the turtle is told to move in a certain way it can leave behind a tracing in the shape of a square, triangle, circle, or almost any other geometric shape. This, however, is not the only way to draw with the turtle. Into most versions of LOGO there is built an (x,y) plane, so that the turtle can be sent to different places by specifying the coordinates of its next location. When the pen is down, a line can be drawn from one point to the next, and geometric figures can be drawn in this way. When both of the two methods of drawing in LOGO are used together, a powerful drawing capability is unleashed.

LOGO also has a unique system of making programs. Instead of each program being a separate pattern for the computer to follow, the user is allowed to design his or her own commands. The basic commands, such as PRINT, FORWARD, HOME, and MAKE are already defined by the computer, and they are called "primitives." Each procedure that the *user* creates is like expanding the LOGO vocabulary to include a *new* function. After defining the procedure under a certain name, the user needs only to type that name to activate the procedure and have the computer carry out its functions. Each procedure can, in turn, contain the name of another procedure, which will activate it, and so on. The user can create a network of procedures, with each calling another into action, when necessary.

Procedures can even activate themselves, going into a cycle. Each procedure begins with the word "TO" followed by the name, and terminates with the word "END." Most of the times that numbers need to be entered for the program to work, they can be typed immediately after the name of the program. These are called "arguments," and are set up after the program's name by typing a colon followed by the name of a variable assigned to the number entered.

Please note that this CP page is by no means an explanation of the LOGO language. It is simply intended to give a beginner a basic idea of what LOGO is. It will usually be advisable to have read a LOGO manual and know your way around the primitives well before working with the pages.

When you type in the programs

When entering the programs, if a line is indented from the margin, it means that it is a continuation of the previous line, with a space between the last word of the last line and the first word of the indented line. If a procedure is cut off at the bottom of a page and continued on the next, keep typing as if nothing had happened. Remember that the only things dividing programs are the "TO" and "END" commands.

Using procedures from one page on another page

As you begin to type in the procedures on various pages, you may notice that several procedures are commonly used and are repeated on different pages. First, this means that if a procedure appears on one page but has no explanation at the end of the page, the procedure must have appeared previously on a different page. Second, this means that you *might* not have to type in the repeated procedure(s) again if you've already done it once. For example: the CP3 page uses many of the same procedures as the CP2 page. If you typed in all of the procedures on the CP2 page and saved them on a disk, then instead of typing in the repeated ones again when it was time to do CP3, you could simply load up and define the CP2 procedures from disk. If you did this, you would only have to type in the procedures on CP3 that do not appear on CP2, reducing the amount of typing significantly. At the very end of each CP-page is a paragraph containing a list of pages which contain many of the repeated procedures on that page, and if you've already typed them in earlier, you might as well load them in (so LOGO knows how to perform those procedures) and save some typing. If you do this, the paragraph also tells which brand new procedures from the page still need to be typed in. All this process does, however, is save a little time, and things certainly don't need to be done in this way. Each and every page has been written so that it stands alone, and everything will work fine if you simply type in all of the procedures on that page.

Different versions of LOGO

All of the procedures, as they appear in the following pages, are written to work perfectly on LCSI LOGO II for Apple II computers. Below, lists of necessary modifications are given *if* you plan to use these programs with LCSI Logowriter (version 2.03) for Apple II computers or Terrapin Logo for Macintosh Computers. If you plan to use these programs with still *other* versions of LOGO, you

will need to consult your reference manual for the necessary
modifications. Since LOGO is a relatively universal language, the
number of modifications should never be large.*

Remember that if you are using LCSI LOGO II, no modifications
whatsoever are needed.

When using Terrapin Logo for the Macintosh: Wherever a
procedure by the same name as one of the following appears, replace
it with the version of the procedure in the list below:

```
TO TURNOFFWRAP
HT NOWRAP
END

TO SPLIT
END

TO TEXTVIEW
END

TO HEADTOWARDS :PAIR
OP TOWARDSPOS :PAIR
END
```

In addition, add this procedure to all pages' lists of procedures:

```
TO CS
CG
END
```

When using LCSI Logowriter for Apple II: Wherever a procedure
by the same name as one of the following appears, replace it with
the version of the procedure in the list below:

```
TO TURNOFFWRAP
HT IF NOT FRONT? [FLIP]
END

TO TEXTVIEW
CT
END

TO FINDRAYEND :PT1 :PT2
MAKE "1STPARAM T.RECTINTERSECT :PT1 :PT2 85 85
MAKE "2NDPARAM T.RECTINTERSECT :PT1 :PT2 -85 -85
IF (ABSVAL :1STPARAM) < (ABSVAL :2NDPARAM) [OUTPUT PARAMLINEPT :PT1 :PT2
     :1STPARAM]
OUTPUT PARAMLINEPT :PT1 :PT2 :2NDPARAM]
END

TO RAYEND :PT1 :PT2
OP FINDRAYEND :PT1 :PT2
END
```

Logowriter, continued: Also, you'll need to add the following
procedures to all pages requiring FINDRAYEND:

```
TO T.INTERSECT :PNUM :QNUM :BOUND
IF :PNUM > :QNUM [MAKE "BOUND -1 * :BOUND]
OP (:BOUND - :PNUM) / (:QNUM - :PNUM)
END

TO T.RECTINTERSECT :PTP :PTQ :XMAX :YMAX
IF (ABSVAL T.INTERSECT FIRST :PTP FIRST :PTQ :XMAX) < (ABSVAL T.INTERSECT
    LAST :PTP LAST :PTQ :YMAX) [OP T.INTERSECT FIRST :PTP FIRST :PTQ :XMAX]
OP T.INTERSECT LAST :PTP LAST :PTQ :YMAX
END

TO ABSVAL :NUM
IF :NUM < Ø [OP -1 * :NUM]
OP :NUM
END
```

Lastly, you'll need to add these procedures to all pages' lists of
procedures:

```
TO CLEARTEXT
CT
END

TO CS
CG
END

TO LOCAL :X
END

TO RC
OP READCHAR
END
```

With these modifications, all of the CP-pages should work without
any problems. If you do receive an error message from LOGO while
attempting to use any of the programs, the most likely problem is
that a typing mistake was made while the program was being put into
the computer. An additional cause of errors is overstepping the
input boundaries of the programs, for example entering (150,299) as
a point on CP2 or entering the same point for both inputs on CP4.
Notice that its a good idea to keep coordinates, magnification
factors, translations, etc. to small numbers. After all, most of
the display grids are only 10x10 or 15x15 units, and if a point's
coordinates are much larger, then either the program will function
improperly or the point simply won't be seen.

Given two points, construct the segment, ray, and line that pass through them

This page uses concepts from I1 and P2.

Instructions: To activate this set of procedures, type CP2 followed by the two points. You must enter the points in brackets, not parentheses, with the coordinates separated by a space, not a comma. The computer will draw the segment, and then stop until you hit any key. It will do the same for the ray, and finish with the line.

Example: To construct the segment, ray, and line through (-2,-6) and (7,2), type: CP2 [-2 -6] [7 2]

    After drawing the segment and ray, the computer would draw the line as follows:

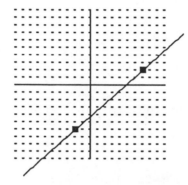

All the procedures needed to carry out the above function are listed below. They must be typed into the computer exactly as shown. Remember: any line that is indented from the margin is a continuation of the previous line.

```
TO CP2 :PT1 :PT2
TURNOFFWRAP
SPLIT CLEARTEXT
DRAWCOORDS
MAKE "PT1 ADJUSTPOINT :PT1
MAKE "PT2 ADJUSTPOINT :PT2
DRAW.SEG.RAY.LINE :PT1 :PT2
END

TO TURNOFFWRAP
HT WINDOW
END

TO DRAW.SEG.RAY.LINE :PT1 :PT2
DRAWPT :PT1 DRAWPT :PT2
DRAWSEG :PT1 :PT2
PR [>>>SEGMENT<<<]
WAITKEY CLEARTEXT
DRAWRAY :PT1 :PT2
PR [>>>RAY<<<]
WAITKEY CLEARTEXT
DRAWLINE :PT1 :PT2
PR [>>>LINE<<<]
END
```

```
TO SPROD :SCALAR :PAIR
OP LIST :SCALAR * (FIRST :PAIR) :SCALAR * (LAST :PAIR)
END


TO DRAWGRID
PD REPEAT 4 [FD 6Ø BK 6Ø RT 9Ø] PU
REPEAT 2 [RT 9Ø FD 6Ø] RT 9Ø
REPEAT 21 [REPEAT 21 [PD FD 1 PU FD 5] BK 126 RT 9Ø FD 6 LT 9Ø]
PU HOME
END


TO DRAWCOORDS
HOME CS
DRAWGRID
END


TO DRAWPT :PT
PU SETPOS :PT PD
MAKE "OLDHEAD HEADING
CENTERSQ 1
CENTERSQ 2
SETH :OLDHEAD
END


TO CENTERSQ :RADIUS
PU FD :RADIUS RT 9Ø
PD REPEAT 4 [FD :RADIUS RT 9Ø FD :RADIUS]
PU LT 9Ø BK :RADIUS
END


TO FDIST :PTX :PTY
OP HYP (FIRST :PTX) - (FIRST :PTY) (LAST :PTX) - (LAST :PTY)
END


TO HYP :SIDEA :SIDEB
OP SQRT (:SIDEA * :SIDEA) +  (:SIDEB * :SIDEB)
END


TO PARAMLINEPT :PT1 :PT2 :PARAM
OP LIST PARAMNUM FIRST :PT1 FIRST :PT2 :PARAM PARAMNUM LAST :PT1 LAST :PT2
    :PARAM
END


TO PARAMNUM :NUM1 :NUM2 :PARAM
OP :NUM1 + ((:NUM2 - :NUM1) * :PARAM)
END


TO RAYEND :PT1 :PT2
OP FINDRAYEND :PT1 :PT2 16Ø Ø
END


TO FINDRAYEND :PT1 :PT2 :MAXDIST :PARAM
LOCAL "NEWRAYEND
MAKE "NEWRAYEND PARAMLINEPT :PT1 :PT2 :PARAM
IF 16Ø < FDIST [Ø Ø] :NEWRAYEND [OP :NEWRAYEND]
OP FINDRAYEND :PT1 :PT2 :MAXDIST :PARAM + 1
END


TO DRAWSEG :PT1 :PT2
PU SETPOS :PT1
PD SETPOS :PT2
END
```

```
TO DRAWRAY :PT1 :PT2
DRAWSEG :PT1 RAYEND :PT1 :PT2
END

TO DRAWLINE :PT1 :PT2
DRAWSEG RAYEND :PT2 :PT1 RAYEND :PT1 :PT2
END

TO ADJUSTPOINT :P
OP SPROD 6 :P
END

TO WAITKEY
PR [PRESS ANY KEY TO CONTINUE]
IGNORE RC
END

TO IGNORE :X
END

TO SPLIT
SS
END
```

The purpose of each of the above procedures is summarized below:

CP2 activates TURNOFFWRAP, DRAWCOORDS, and DRAW.SEG.RAY.LINE.

TURNOFFWRAP hides the turtle and turns off the WRAP feature that
    causes the turtle to leave one side of the screen and reappear on
    the other side.

DRAW.SEG.RAY.LINE begins by running the two points through ADJUSTPOINT.
    It then activates SPLIT, clears all text, and draws the points and
    the segment. It then labels the segments as such, and waits for
    a key to be hit before continuing to draw the ray and label it.
    After another key hit, it draws the line.

SPROD, short for Scalar PRODuct, multiplies both numbers within a
    vector or coordinate pair by a scalar.

DRAWGRID draws the x-axis and y-axis, then creates a 10x10 unit grid
    that is 60x60 in turtle steps.

DRAWCOORDS centers the turtle, clears the screen, and activates
    DRAWGRID.

DRAWPT draws a coordinate pair on the grid by placing the turtle at
    the point's position and drawing a CENTERSQ.

FDIST uses the Distance Formula to find the distance between two
    points. It outputs the HYP of the triangle whose two legs are
    determined by the two points.

HYP uses the Pythagorean Theorem to output the HYPoteneuse of a
    triangle with legs the length of its two inputs.

PARAMLINEPT outputs the coordinates of a new ray-end to FINDRAYEND by
    using PARAMNUM.

PARAMNUM adds another multiple of either the x or y component of the
    vector between the two points from PARAMLINEPT, depending on the
    size of :PARAM.

RAYEND tells FINDRAYEND to find a ray-end that is at least 160 turtle
    steps away from the origin (0,0).

FINDRAYEND uses PARAMLINEPT to add another copy of the vector between
    the two points onto the ray-end it is working on, then checks to
    see if this new end is far enough from the origin. If not, it
    repeats with a larger :PARAM.

DRAWSEG moves the turtle from one point to another with the pen down.
DRAWRAY draws a segment from one point to the RAYEND that is past the
    other point.
DRAWLINE draws a segment from the RAYEND that is past one point to the
    RAYEND that is past the other point.
ADJUSTPOINT outputs the scalar product of the input point and 6.
    This aligns the point to the enlarged grid from DRAWGRID.
WAITKEY reads character (RC) from the keyboard before going any
    further. IGNORE does nothing, so the input character is ignored.
SPLIT splits the screen between text and graphics.

Note: In LOGO, variable names that are written right after the name
of a procedure (these variables are called "arguments") are local to
that procedure, not global throughout the entire program. For
example, in the line MAKE "PT1 ADJUSTPOINT :PT1, the value of :PT1, not
the name, is passed to ADJUSTPOINT. Upon reaching ADJUSTPOINT, it is
temporarily renamed :P and is dealt with by that name while inside
ADJUSTPOINT. This is important to understand if you plan to use
procedures from different CP sections together to make your own
programs. For example, when CP8 used the DRAW.PERP.OR.PARA procedure
from CP3, it was not necessary to rename the :LINEPT1 variable within
that procedure to match the :PT1 variable name from the main CP8.
This is true because when DRAW.ALTITUDES calls DRAW.PERP.OR.PARA, it
passes the values of :PT1 etc., not the names. So the value of :PT1,
for instance, is automatically assigned the name of :LINEPT1 upon
reaching that procedure. Remember, however, that this is only true
of the "argument" variables that follow procedure names.

Given a line and a point, construct the perpendicular to the line through the point, or the parallel to the line through the point

This page uses concepts from I14, C3, and C4.

Instructions: To activate this set of procedures, type CP3, followed by the two points that determine the line, followed by the point through which the perpendicular or parallel will pass, followed by either "PERP or "PARA.

Example: If the given line passes through (-5,1) and (7,-5), then to construct the unique perpendicular to this line that passes through (2,6), type: CP3 [-5 1] [7 -5] [2 6] "PERP
   The computer will draw:

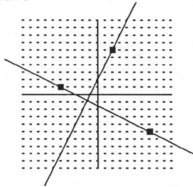

All of the procedures needed to carry out the above functions are listed below. They must be typed into the computer exactly as shown. Remember: any line that is indented from the margin is a continuation of the previous line.


```
TO CP3 :LINEPT1 :LINEPT2 :POINT :PERPORPARA
TURNOFFWRAP
MAKE "LINEPT1 ADJUSTPOINT :LINEPT1
MAKE "LINEPT2 ADJUSTPOINT :LINEPT2
MAKE "POINT ADJUSTPOINT :POINT
SPLIT CLEARTEXT DRAWCOORDS
DRAW.LINE.POINT :LINEPT1 :LINEPT2 :POINT
DRAW.PERP.OR.PARA :LINEPT1 :LINEPT2 :POINT :PERPORPARA
END

TO VECTOR :P :Q
OP LIST (FIRST :Q) - (FIRST :P) (LAST :Q) - (LAST :P)
END

TO PERP :VECT
OP LIST -1 * LAST :VECT FIRST :VECT
END

TO PARA :VECT
OP :VECT
END
```

```
TO SUM.POINT.VECT :P :Q
OP LIST (FIRST :P) + (FIRST :Q) (LAST :P) + (LAST :Q)
END


TO DRAW.PERP.OR.PARA :LINEPT1 :LINEPT2 :POINT :CHOICE
IF :CHOICE = "PERP [DRAWRESULT :POINT PERP VECTOR :LINEPT1 :LINEPT2 STOP]
DRAWRESULT :POINT PARA VECTOR :LINEPT1 :LINEPT2
END


TO DRAWRESULT :POINT :VECT1
DRAWLINE :POINT SUM.POINT.VECT :POINT :VECT1
END


TO DRAW.LINE.POINT :LINEPT1 :LINEPT2 :POINT
DRAWPT :LINEPT1
DRAWPT :LINEPT2
DRAWPT :POINT
DRAWLINE :LINEPT1 :LINEPT2
END


TO FDIST :PTX :PTY
OP HYP (FIRST :PTX) - (FIRST :PTY) (LAST :PTX) - (LAST :PTY)
END


TO HYP :SIDEA :SIDEB
OP SQRT (:SIDEA * :SIDEA) +  (:SIDEB * :SIDEB)
END


TO TURNOFFWRAP
HT WINDOW
END


TO SPLIT
SS
END


TO DRAWCOORDS
HOME CS
DRAWGRID
END


TO DRAWGRID
PD REPEAT 4 [FD 60 BK 60 RT 90] PU
REPEAT 2 [RT 90 FD 60] RT 90
REPEAT 21 [REPEAT 21 [PD FD 1 PU FD 5] BK 126 RT 90 FD 6 LT 90]
PU HOME
END


TO DRAWPT :PT
PU SETPOS :PT PD
MAKE "OLDHEAD HEADING
CENTERSQ 1
CENTERSQ 2
SETH :OLDHEAD
END


TO CENTERSQ :RADIUS
PU FD :RADIUS RT 90
PD REPEAT 4 [FD :RADIUS RT 90 FD :RADIUS]
PU LT 90 BK :RADIUS
END
```

```
TO DRAWLINE :PT1 :PT2
DRAWSEG RAYEND :PT2 :PT1 RAYEND :PT1 :PT2
END

TO DRAWSEG :PT1 :PT2
PU SETPOS :PT1
PD SETPOS :PT2
END

TO RAYEND :PT1 :PT2
OP FINDRAYEND :PT1 :PT2 16Ø Ø
END

TO FINDRAYEND :PT1 :PT2 :MAXDIST :PARAM
LOCAL "NEWRAYEND
MAKE "NEWRAYEND PARAMLINEPT :PT1 :PT2 :PARAM
IF 16Ø < FDIST [Ø Ø] :NEWRAYEND [OP :NEWRAYEND]
OP FINDRAYEND :PT1 :PT2 :MAXDIST :PARAM + 1
END

TO PARAMLINEPT :PT1 :PT2 :PARAM
OP LIST PARAMNUM FIRST :PT1 FIRST :PT2 :PARAM PARAMNUM LAST :PT1 LAST :PT2
   :PARAM
END

TO PARAMNUM :NUM1 :NUM2 :PARAM
OP :NUM1 + ((:NUM2 - :NUM1) * :PARAM)
END

TO ADJUSTPOINT :P
OP SPROD 6 :P
END

TO SPROD :SCALAR :PAIR
OP LIST :SCALAR * (FIRST :PAIR) :SCALAR * (LAST :PAIR)
END
```

The purpose of each of the above procedures is summarized below, if it has not been summarized on a previous page.

CP3 activates TURNOFFWRAP, runs the entered points through ADJUSTPOINT, activates DRAW.LINE.POINT and DRAW.PERP.OR.PARA.
VECTOR outputs a pair of numbers that represents the vector from the first input point to the second.
PERP accepts a vector as input and outputs a vector that is perpendicular to the input.
PARA accepts a vector as input and outputs a vector that is parallel to the input (the same vector).
SUM.POINT.VECT moves its first input, a coordinate pair, by adding to it a vector, also a pair of numbers.
DRAW.PERP.OR.PARA picks what to output to DRAWRESULT depending on whether the :CHOICE is "PERP or "PARA.
DRAWRESULT draws a line through the input point and a point whose difference from the point is the vector :VECT1.
DRAW.LINE.POINT activates SPLIT and DRAWCOORDS, and then draws the input points and input line from CP3.

<u>Please note</u>: This CP-page *does* stand alone, and all of the
procedures necessary to carry out its purpose are listed directly
after the example at the beginning. However, you may notice that
many of the above procedures have appeared previously. If it so
happens that you have already defined (typed in or loaded from disk)
all of the procedures from the CP2 page since you started up LOGO,
then LOGO already knows how to perform many of the above procedures.
If this is the case, the only new procedures from this page that you
need to type in are: CP3, VECTOR, PERP, PARA, SUM.POINT.VECT,
DRAW.PERP.OR.PARA, DRAWRESULT, and DRAW.LINE.POINT.

Given a segment, construct its perpendicular bisector

This page uses concepts from I22e and C3e.

Instructions: To activate this set of procedures, type CP4 followed
by the two points that determine your segment. You must enter the
points in brackets, not parentheses, with the coordinates separated
by a space, not a comma.

Example: To construct the perpendicular bisector of the segment
from (-3,1) to (1,6), type: CP4 [-3 1] [1 -6]
    The computer draws this:

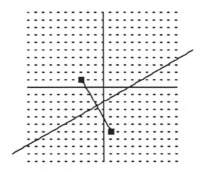

All of the procedures needed to carry out the above functions are
listed below. They must be typed into the computer exactly as
shown. Remember: any line that is indented from the margin is a
continuation of the previous line.

```
TO CP4 :PT1 :PT2
TURNOFFWRAP
SPLIT CLEARTEXT
DRAWCOORDS
MAKE "PT1 ADJUSTPOINT :PT1
MAKE "PT2 ADJUSTPOINT :PT2
DRAW.BISECTOR :PT1 :PT2
END


TO DRAW.BISECTOR :PT1 :PT2
DRAWPT :PT1
DRAWPT :PT2
DRAWSEG :PT1 :PT2
DRAWLINE MIDPOINT :PT1 :PT2 SUM.POINT.VECT MIDPOINT :PT1 :PT2 PERP VECTOR :PT1
    :PT2
END


TO MIDPOINT :PT1 :PT2
OP LIST MIDCOORD FIRST :PT1 FIRST :PT2 MIDCOORD LAST :PT1 LAST :PT2
END


TO MIDCOORD :P :Q
OP (:P + :Q) / 2
END
```

```
TO FDIST :PTX :PTY
OP HYP (FIRST :PTX) - (FIRST :PTY) (LAST :PTX) - (LAST :PTY)
END


TO HYP :SIDEA :SIDEB
OP SQRT (:SIDEA * :SIDEA) +  (:SIDEB * :SIDEB)
END


TO TURNOFFWRAP
HT WINDOW
END


TO SPLIT
SS
END


TO DRAWCOORDS
HOME CS
DRAWGRID
END


TO DRAWGRID
PD REPEAT 4 [FD 6Ø BK 6Ø RT 9Ø] PU
REPEAT 2 [RT 9Ø FD 6Ø] RT 9Ø
REPEAT 21 [REPEAT 21 [PD FD 1 PU FD 5] BK 126 RT 9Ø FD 6 LT 9Ø]
PU HOME
END


TO DRAWPT :PT
PU SETPOS :PT PD
MAKE "OLDHEAD HEADING
CENTERSQ 1
CENTERSQ 2
SETH :OLDHEAD
END


TO CENTERSQ :RADIUS
PU FD :RADIUS RT 9Ø
PD REPEAT 4 [FD :RADIUS RT 9Ø FD :RADIUS]
PU LT 9Ø BK :RADIUS
END


TO DRAWLINE :PT1 :PT2
DRAWSEG RAYEND :PT2 :PT1 RAYEND :PT1 :PT2
END


TO DRAWSEG :PT1 :PT2
PU SETPOS :PT1
PD SETPOS :PT2
END


TO RAYEND :PT1 :PT2
OP FINDRAYEND :PT1 :PT2 16Ø Ø
END


TO FINDRAYEND :PT1 :PT2 :MAXDIST :PARAM
LOCAL "NEWRAYEND
MAKE "NEWRAYEND PARAMLINEPT :PT1 :PT2 :PARAM
IF 16Ø < FDIST [Ø Ø] :NEWRAYEND [OP :NEWRAYEND]
OP FINDRAYEND :PT1 :PT2 :MAXDIST :PARAM + 1
END
```

```
TO PARAMLINEPT :PT1 :PT2 :PARAM
OP LIST PARAMNUM FIRST :PT1 FIRST :PT2 :PARAM PARAMNUM LAST :PT1 LAST :PT2
   :PARAM
END

TO PARAMNUM :NUM1 :NUM2 :PARAM
OP :NUM1 + ((:NUM2 - :NUM1) * :PARAM)
END

TO ADJUSTPOINT :P
OP SPROD 6 :P
END

TO SPROD :SCALAR :PAIR
OP LIST :SCALAR * (FIRST :PAIR) :SCALAR * (LAST :PAIR)
END

TO VECTOR :P :Q
OP LIST (FIRST :Q) - (FIRST :P) (LAST :Q) - (LAST :P)
END

TO PERP :VECT
OP LIST -1 * LAST :VECT FIRST :VECT
END

TO SUM.POINT.VECT :P :Q
OP LIST (FIRST :P) + (FIRST :Q) (LAST :P) + (LAST :Q)
END
```

The purpose of each of the above procedures is summarized below, if it has not been summarized on a previous page.

CP4 activates TURNOFFWRAP, SPLIT, DRAWCOORDS, and clears all text.
    After running both points through ADJUSTPOINT, it moves on to
    DRAW.BISECTOR.
DRAW.BISECTOR draws the points and segment, then draws a line that
    passes through 1) the midpoint of the segment and 2) a point
    whose offset from the midpoint is a vector perpendicular to the
    vector from the first entered point to the second.
MIDPOINT outputs a point whose coordinates are figured from the
    MIDCOORDs of the entered points. It uses the formula from I22e to
    figure the midpoint.
MIDCOORD outputs the average of the two inputs.

Please note: This CP-page *does* stand alone, and all of the
procedures necessary to carry out its purpose are listed directly
after the example at the beginning. However, you may notice that
many of the above procedures have appeared previously. If it so
happens that you have already defined (typed in or loaded from disk)
all of the procedures from the CP2 page and the CP3 page since you
started up LOGO, then LOGO already knows how to perform many of the
above procedures. If this is the case, the only new procedures from
this page that you need to type in are: CP4, DRAW.BISECTOR, MIDPOINT,
and MIDCOORD.

Given an angle, construct the bisector

This page uses concepts from C18.

<u>Instructions</u>: To activate this set of procedures, type CP5 followed by the coordinates of the vertex of the angle, followed by the coordinates of two points, each of which lies on one ray of the angle. You must enter the points in brackets, not parentheses, with the coordinates separated by a space, not a comma.

<u>Example</u>: To bisect the angle with vertex (4,-5) and whose rays contain the points (-2,2) and (8,5), type: CP5 [4 -5] [-2 2] [8 5]
     The computer draws this:

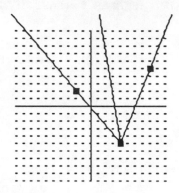

All of the procedures needed to carry out the above functions are listed below. They must be typed into the computer exactly as shown. Remember: any line that is indented from the margin is a continuation of the previous line.

```
TO CP5 :VTX :PT1 :PT2
TURNOFFWRAP
SPLIT CLEARTEXT
DRAWCOORDS
MAKE "VTX ADJUSTPOINT :VTX
MAKE "PT1 ADJUSTPOINT :PT1
MAKE "PT2 ADJUSTPOINT :PT2
DRAW.ANGLE.BISECTOR :VTX :PT1 :PT2
END


TO DRAW.ANGLE.BISECTOR :VTX :PT1 :PT2
DRAWPT :VTX
DRAWPT :PT1
DRAWPT :PT2
DRAWRAY :VTX :PT1
DRAWRAY :VTX :PT2
DRAWRAY :VTX PT.ON.ANG.BISECT :VTX :PT1 :PT2 3Ø
END


TO PT.ON.ANG.BISECT :VTX :PT1 :PT2 :DIST
LOCAL "VECT1EQDIST
LOCAL "VECT2EQDIST
MAKE "VECT1EQDIST SPROD (:DIST/(FDIST :VTX :PT1)) VECTOR :VTX :PT1
MAKE "VECT2EQDIST SPROD (:DIST/(FDIST :VTX :PT2)) VECTOR :VTX :PT2
OP MIDPOINT SUM.POINT.VECT :VTX :VECT1EQDIST SUM.POINT.VECT :VTX :VECT2EQDIST
END
```

```
TO FDIST :PTX :PTY
OP HYP (FIRST :PTX) - (FIRST :PTY) (LAST :PTX) - (LAST :PTY)
END

TO HYP :SIDEA :SIDEB
OP SQRT (:SIDEA * :SIDEA) +  (:SIDEB * :SIDEB)
END

TO TURNOFFWRAP
HT WINDOW
END

TO SPLIT
SS
END

TO DRAWCOORDS
HOME CS
DRAWGRID
END

TO DRAWGRID
PD REPEAT 4 [FD 6Ø BK 6Ø RT 9Ø] PU
REPEAT 2 [RT 9Ø FD 6Ø] RT 9Ø
REPEAT 21 [REPEAT 21 [PD FD 1 PU FD 5] BK 126 RT 9Ø FD 6 LT 9Ø]
PU HOME
END

TO DRAWPT :PT
PU SETPOS :PT PD
MAKE "OLDHEAD HEADING
CENTERSQ 1
CENTERSQ 2
SETH :OLDHEAD
END

TO CENTERSQ :RADIUS
PU FD :RADIUS RT 9Ø
PD REPEAT 4 [FD :RADIUS RT 9Ø FD :RADIUS]
PU LT 9Ø BK :RADIUS
END

TO DRAWRAY :PT1 :PT2
DRAWSEG :PT1 RAYEND :PT1 :PT2
END

TO DRAWSEG :PT1 :PT2
PU SETPOS :PT1
PD SETPOS :PT2
END

TO RAYEND :PT1 :PT2
OP FINDRAYEND :PT1 :PT2 16Ø Ø
END

TO FINDRAYEND :PT1 :PT2 :MAXDIST :PARAM
LOCAL "NEWRAYEND
MAKE "NEWRAYEND PARAMLINEPT :PT1 :PT2 :PARAM
IF 16Ø < FDIST [Ø Ø] :NEWRAYEND [OP :NEWRAYEND]
OP FINDRAYEND :PT1 :PT2 :MAXDIST :PARAM + 1
END
```

```
TO PARAMLINEPT :PT1 :PT2 :PARAM
OP LIST PARAMNUM FIRST :PT1 FIRST :PT2 :PARAM PARAMNUM LAST :PT1 LAST :PT2
   :PARAM
END

TO PARAMNUM :NUM1 :NUM2 :PARAM
OP :NUM1 + ((:NUM2 - :NUM1) * :PARAM)
END

TO ADJUSTPOINT :P
OP SPROD 6 :P
END

TO SPROD :SCALAR :PAIR
OP LIST :SCALAR * (FIRST :PAIR) :SCALAR * (LAST :PAIR)
END

TO VECTOR :P :Q
OP LIST (FIRST :Q) - (FIRST :P) (LAST :Q) - (LAST :P)
END

TO SUM.POINT.VECT :P :Q
OP LIST (FIRST :P) + (FIRST :Q) (LAST :P) + (LAST :Q)
END

TO MIDPOINT :PT1 :PT2
OP LIST MIDCOORD FIRST :PT1 FIRST :PT2 MIDCOORD LAST :PT1 LAST :PT2
END

TO MIDCOORD :P :Q
OP (:P + :Q) / 2
END
```

The purpose of each of the above procedures is summarized below, if it has not been summarized on a previous page.

CP5 activates TURNOFFWRAP, SPLIT, DRAWCOORDS, and clears all text. It then moves on to DRAW.ANGLE.BISECTOR.

DRAW.ANGLE.BISECTOR runs all points through ADJUSTPOINT, draws the points, draw the rays of the angle. It then draws a ray from the vertex to a point on the angle bisector, as determined by PT.ON.ANG.BISECT.

PT.ON.ANG.BISECT outputs the coordinates of a point on the angle bisector. First, it finds the vector from the vertex to each of the other input points, the points on the rays. By dividing each component of each vector by the point's distance from the vertex, it finds the vector from the vertex towards that point whose length is 1. It then multiplies each unit vector by :DIST to give it significant length. This results in two vectors, :VECT1EQDIST and :VECT2EQDIST which represent vectors along each ray that are the same length, :DIST. When these are added to the vertex coordinates with SUM.POINT.VECT, the result is the coordinates of two points, one on each ray, both of which are equidistant from the vertex. The angle bisector will pass through the midpoint of these two resulting points. This is *exactly* what is done on C18.

<u>Please note</u>: This CP-page *does* stand alone, and all of the
procedures necessary to carry out its purpose are listed directly
after the example at the beginning. However, you may notice that
many of the above procedures have appeared previously. If it so
happens that you have already defined (typed in or loaded from disk)
all of the procedures from the CP2 page, the CP3 page, and the CP4
page since you started up LOGO, then LOGO already knows how to
perform many of the above procedures. If this is the case, the only
new procedures from this page that you need to type in are: CP5,
DRAW.ANGLE.BISECTOR, and PT.ON.ANG.BISECT.

Given three vertices, construct the triangle and its medians

This page uses concepts from I22 and C15.

<u>Instructions</u>: To activate this set of procedures, type CP6 followed by the coordinates of the three vertices. You must enter the points in brackets, not parentheses, with the coordinates separated by a space, not a comma.

<u>Example</u>: To construct the medians of the triangle whose vertices are (-8,5), (-10 -7), and (2,-5), type: CP6 [-8 5] [-10 -7] [2 -5]
     The computer will draw:

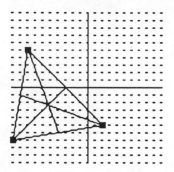

All of the procedures needed to carry out the above functions are listed below. They must be typed into the computer exactly as shown. Remember: any line that is indented from the margin is a continuation of the previous line.

```
TO CP6 :PT1 :PT2 :PT3
TURNOFFWRAP
SPLIT CLEARTEXT
DRAWCOORDS
MAKE "PT1 ADJUSTPOINT :PT1
MAKE "PT2 ADJUSTPOINT :PT2
MAKE "PT3 ADJUSTPOINT :PT3
DRAW.TRI.MEDIANS :PT1 :PT2 :PT3
END

TO DRAW.TRI.MEDIANS :PT1 :PT2 :PT3
DRAWTRI :PT1 :PT2 :PT3
DRAW.MEDIANS :PT1 :PT2 :PT3
END

TO DRAWTRI :PT1 :PT2 :PT3
DRAWSEG :PT1 :PT2
DRAWSEG :PT2 :PT3
DRAWSEG :PT3 :PT1
DRAWPT :PT1
DRAWPT :PT2
DRAWPT :PT3
END
```

```
TO DRAW.MEDIANS :PT1 :PT2 :PT3
DRAWSEG :PT1 MIDPOINT :PT2 :PT3
DRAWSEG :PT2 MIDPOINT :PT1 :PT3
DRAWSEG :PT3 MIDPOINT :PT1 :PT2
END

TO MIDPOINT :PT1 :PT2
OP LIST MIDCOORD FIRST :PT1 FIRST :PT2 MIDCOORD LAST :PT1 LAST :PT2
END

TO MIDCOORD :P :Q
OP (:P + :Q) / 2
END

TO TURNOFFWRAP
HT WINDOW
END

TO SPLIT
SS
END

TO DRAWCOORDS
HOME CS
DRAWGRID
END

TO DRAWGRID
PD REPEAT 4 [FD 6Ø BK 6Ø RT 9Ø] PU
REPEAT 2 [RT 9Ø FD 6Ø] RT 9Ø
REPEAT 21 [REPEAT 21 [PD FD 1 PU FD 5] BK 126 RT 9Ø FD 6 LT 9Ø]
PU HOME
END

TO DRAWPT :PT
PU SETPOS :PT PD
MAKE "OLDHEAD HEADING
CENTERSQ 1
CENTERSQ 2
SETH :OLDHEAD
END

TO CENTERSQ :RADIUS
PU FD :RADIUS RT 9Ø
PD REPEAT 4 [FD :RADIUS RT 9Ø FD :RADIUS]
PU LT 9Ø BK :RADIUS
END

TO DRAWSEG :PT1 :PT2
PU SETPOS :PT1
PD SETPOS :PT2
END

TO ADJUSTPOINT :P
OP SPROD 6 :P
END

TO SPROD :SCALAR :PAIR
OP LIST :SCALAR * (FIRST :PAIR) :SCALAR * (LAST :PAIR)
END
```

The purpose of each of the above procedures is summarized below, if it has not been summarized on a previous page.

CP6 activates TURNOFFWRAP, SPLIT, DRAWCOORDS, and clears all text. It
    then moves on to DRAW.TRI.MEDIANS.
DRAW.TRI.MEDIANS runs all points through ADJUSTPOINT, then moves on to
    DRAWTRI and DRAW.MEDIANS.
DRAWTRI draws the entered points and the segments between each pair.
DRAW.MEDIANS draws a segment from each vertex to the midpoint of the
    opposite side.

Please note: This CP-page *does* stand alone, and all of the
procedures necessary to carry out its purpose are listed directly
after the example at the beginning. However, you may notice that
many of the above procedures have appeared previously. If it so
happens that you have already defined (typed in or loaded from disk)
all of the procedures from the CP2 page and the CP4 page since you
started up LOGO, then LOGO already knows how to perform many of the
above procedures. If this is the case, the only new procedures from
this page that you need to type in are: CP6, DRAW.TRI.MEDIANS, DRAWTRI,
and DRAW.MEDIANS.

Given three vertices, construct the triangle and its angle bisectors

This page uses concepts from C18 and C18e.

Instructions: To activate this set of procedures, type CP7 followed by the coordinates of the three vertices. You must enter the points in brackets, not parentheses, with the coordinates separated by a space, not a comma.

Example: To construct the angle bisectors of the triangle with vertices (-3,3), (8,-2), and (9,5), type: CP7 [-3 3] [8 -2] [9 5]
    The computer will draw:

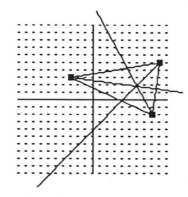

All of the procedures needed to carry out the above functions are listed below. They must be typed into the computer exactly as shown. Remember: any line that is indented from the margin is a continuation of the previous line.

```
TO CP7 :PT1 :PT2 :PT3
TURNOFFWRAP
SPLIT CLEARTEXT
DRAWCOORDS
MAKE "PT1 ADJUSTPOINT :PT1
MAKE "PT2 ADJUSTPOINT :PT2
MAKE "PT3 ADJUSTPOINT :PT3
DRAW.TRI.BISECT :PT1 :PT2 :PT3
END

TO DRAW.TRI.BISECT :PT1 :PT2 :PT3
DRAWTRI :PT1 :PT2 :PT3
DRAW.ANGLE.BISECTS :PT1 :PT2 :PT3
END

TO DRAW.ANGLE.BISECTS :PT1 :PT2 :PT3
DRAWRAY :PT1 PT.ON.ANG.BISECT :PT1 :PT2 :PT3 30
DRAWRAY :PT2 PT.ON.ANG.BISECT :PT2 :PT1 :PT3 30
DRAWRAY :PT3 PT.ON.ANG.BISECT :PT3 :PT1 :PT2 30
END
```

```
TO DRAWTRI :PT1 :PT2 :PT3
DRAWSEG :PT1 :PT2
DRAWSEG :PT2 :PT3
DRAWSEG :PT3 :PT1
DRAWPT :PT1
DRAWPT :PT2
DRAWPT :PT3
END

TO PT.ON.ANG.BISECT :VTX :PT1 :PT2 :DIST
LOCAL "VECT1EQDIST
LOCAL "VECT2EQDIST
MAKE "VECT1EQDIST SPROD (:DIST/(FDIST :VTX :PT1)) VECTOR :VTX :PT1
MAKE "VECT2EQDIST SPROD (:DIST/(FDIST :VTX :PT2)) VECTOR :VTX :PT2
OP MIDPOINT SUM.POINT.VECT :VTX :VECT1EQDIST SUM.POINT.VECT :VTX :VECT2EQDIST
END

TO FDIST :PTX :PTY
OP HYP (FIRST :PTX) - (FIRST :PTY) (LAST :PTX) - (LAST :PTY)
END

TO HYP :SIDEA :SIDEB
OP SQRT (:SIDEA * :SIDEA) +  (:SIDEB * :SIDEB)
END

TO TURNOFFWRAP
HT WINDOW
END

TO SPLIT
SS
END

TO DRAWCOORDS
HOME CS
DRAWGRID
END

TO DRAWGRID
PD REPEAT 4 [FD 6Ø BK 6Ø RT 9Ø] PU
REPEAT 2 [RT 9Ø FD 6Ø] RT 9Ø
REPEAT 21 [REPEAT 21 [PD FD 1 PU FD 5] BK 126 RT 9Ø FD 6 LT 9Ø]
PU HOME
END

TO DRAWPT :PT
PU SETPOS :PT PD
MAKE "OLDHEAD HEADING
CENTERSQ 1
CENTERSQ 2
SETH :OLDHEAD
END

TO CENTERSQ :RADIUS
PU FD :RADIUS RT 9Ø
PD REPEAT 4 [FD :RADIUS RT 9Ø FD :RADIUS]
PU LT 9Ø BK :RADIUS
END

TO DRAWRAY :PT1 :PT2
DRAWSEG :PT1 RAYEND :PT1 :PT2
END
```

```
TO DRAWSEG :PT1 :PT2
PU SETPOS :PT1
PD SETPOS :PT2
END

TO RAYEND :PT1 :PT2
OP FINDRAYEND :PT1 :PT2 16Ø Ø
END

TO FINDRAYEND :PT1 :PT2 :MAXDIST :PARAM
LOCAL "NEWRAYEND
MAKE "NEWRAYEND PARAMLINEPT :PT1 :PT2 :PARAM
IF 16Ø < FDIST [Ø Ø] :NEWRAYEND [OP :NEWRAYEND]
OP FINDRAYEND :PT1 :PT2 :MAXDIST :PARAM + 1
END

TO PARAMLINEPT :PT1 :PT2 :PARAM
OP LIST PARAMNUM FIRST :PT1 FIRST :PT2 :PARAM PARAMNUM LAST :PT1 LAST :PT2
    :PARAM .
END

TO PARAMNUM :NUM1 :NUM2 :PARAM
OP :NUM1 + ((:NUM2 - :NUM1) * :PARAM)
END

TO ADJUSTPOINT :P
OP SPROD 6 :P
END

TO SPROD :SCALAR :PAIR
OP LIST :SCALAR * (FIRST :PAIR) :SCALAR * (LAST :PAIR)
END

TO VECTOR :P :Q
OP LIST (FIRST :Q) - (FIRST :P) (LAST :Q) - (LAST :P)
END

TO SUM.POINT.VECT :P :Q
OP LIST (FIRST :P) + (FIRST :Q) (LAST :P) + (LAST :Q)
END

TO MIDPOINT :PT1 :PT2
OP LIST MIDCOORD FIRST :PT1 FIRST :PT2 MIDCOORD LAST :PT1 LAST :PT2
END

TO MIDCOORD :P :Q
OP (:P + :Q) / 2
END
```

The purpose of each of the above procedures is summarized below, if
it has not been summarized on a previous page.

CP7 activates TURNOFFWRAP, SPLIT, DRAWCOORDS, and clears all text. It
    then moves on to DRAW.TRI.BISECT.
DRAW.TRI.BISECT runs all points through ADJUSTPOINT, then moves on to
    DRAWTRI and DRAW.ANGLE.BISECTS.
DRAW.ANGLE.BISECTS draws a ray from each vertex through a point on the
    angle bisector for the angle at that vertex, as determined by
    PT.ON.ANG.BISECT.

<u>Please note</u>: This CP-page *does* stand alone, and all of the
procedures necessary to carry out its purpose are listed directly
after the example at the beginning. However, you may notice that
many of the above procedures have appeared previously. *If* it so
happens that you have already defined (typed in or loaded from disk)
all of the procedures from the CP2 page, the CP3 page, the CP4 page,
the CP5 page, and the CP6 page since you started up LOGO, then LOGO
already knows how to perform many of the above procedures. If this
is the case, the only new procedures from this page that you need to
type in are: CP7, DRAW.TRI.BISECT, and DRAW.ANGLE.BISECTS.

Given three vertices, construct the triangle and its altitudes

This page uses concepts from I23 and C16.

<u>Instructions</u>: To activate this set of procedures, type CP8 followed by the coordinates of the three vertices. You must enter the points in brackets, not parentheses, with the coordinates separated by a space, not a comma.

<u>Example</u>: To construct the angle bisectors of the triangle with vertices (-2,-2), (9,-1), and (-6,8), type: CP8 [-2 -2] [9 -1] [-6 8]
   The computer demonstrates that because of the shape of this triangle, the altitudes intersect outside the triangle:

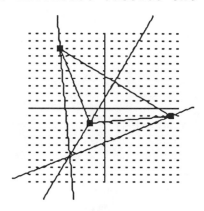

All of the procedures needed to carry out the above functions are listed below. They must be typed into the computer exactly as shown. Remember: any line that is indented from the margin is a continuation of the previous line.

```
TO CP8 :PT1 :PT2 :PT3
TURNOFFWRAP
SPLIT CLEARTEXT
DRAWCOORDS
MAKE "PT1 ADJUSTPOINT :PT1
MAKE "PT2 ADJUSTPOINT :PT2
MAKE "PT3 ADJUSTPOINT :PT3
DRAW.TRI.ALTS :PT1 :PT2 :PT3
END

TO DRAW.TRI.ALTS :PT1 :PT2 :PT3
DRAWTRI :PT1 :PT2 :PT3
DRAW.ALTITUDES :PT1 :PT2 :PT3
END

TO DRAW.ALTITUDES :PT1 :PT2 :PT3
DRAW.PERP.OR.PARA :PT1 :PT2 :PT3 "PERP
DRAW.PERP.OR.PARA :PT2 :PT3 :PT1 "PERP
DRAW.PERP.OR.PARA :PT3 :PT1 :PT2 "PERP
END
```

```
TO DRAWTRI :PT1 :PT2 :PT3
DRAWSEG :PT1 :PT2
DRAWSEG :PT2 :PT3
DRAWSEG :PT3 :PT1
DRAWPT :PT1
DRAWPT :PT2
DRAWPT :PT3
END

TO TURNOFFWRAP
HT WINDOW
END

TO SPLIT
SS
END

TO DRAWCOORDS
HOME CS
DRAWGRID
END

TO DRAWGRID
PD REPEAT 4 [FD 6Ø BK 6Ø RT 9Ø] PU
REPEAT 2 [RT 9Ø FD 6Ø] RT 9Ø
REPEAT 21 [REPEAT 21 [PD FD 1 PU FD 5] BK 126 RT 9Ø FD 6 LT 9Ø]
PU HOME
END

TO DRAWPT :PT
PU SETPOS :PT PD
MAKE "OLDHEAD HEADING
CENTERSQ 1
CENTERSQ 2
SETH :OLDHEAD
END

TO CENTERSQ :RADIUS
PU FD :RADIUS RT 9Ø
PD REPEAT 4 [FD :RADIUS RT 9Ø FD :RADIUS]
PU LT 9Ø BK :RADIUS
END

TO DRAWLINE :PT1 :PT2
DRAWSEG RAYEND :PT2 :PT1 RAYEND :PT1 :PT2
END

TO DRAWSEG :PT1 :PT2
PU SETPOS :PT1
PD SETPOS :PT2
END

TO RAYEND :PT1 :PT2
OP FINDRAYEND :PT1 :PT2 16Ø Ø
END

TO FINDRAYEND :PT1 :PT2 :MAXDIST :PARAM
LOCAL "NEWRAYEND
MAKE "NEWRAYEND PARAMLINEPT :PT1 :PT2 :PARAM
IF 16Ø < FDIST [Ø Ø] :NEWRAYEND [OP :NEWRAYEND]
OP FINDRAYEND :PT1 :PT2 :MAXDIST :PARAM + 1
END
```

```
TO PARAMLINEPT :PT1 :PT2 :PARAM
OP LIST PARAMNUM FIRST :PT1 FIRST :PT2 :PARAM PARAMNUM LAST :PT1 LAST :PT2
   :PARAM
END

TO PARAMNUM :NUM1 :NUM2 :PARAM
OP :NUM1 + ((:NUM2 - :NUM1) * :PARAM)
END

TO ADJUSTPOINT :P
OP SPROD 6 :P
END

TO SPROD :SCALAR :PAIR
OP LIST :SCALAR * (FIRST :PAIR) :SCALAR * (LAST :PAIR)
END

TO VECTOR :P :Q
OP LIST (FIRST :Q) - (FIRST :P) (LAST :Q) - (LAST :P)
END

TO PERP :VECT
OP LIST -1 * LAST :VECT FIRST :VECT
END

TO SUM.POINT.VECT :P :Q
OP LIST (FIRST :P) + (FIRST :Q) (LAST :P) + (LAST :Q)
END

TO DRAW.PERP.OR.PARA :LINEPT1 :LINEPT2 :POINT :CHOICE
IF :CHOICE = "PERP [DRAWRESULT :POINT PERP VECTOR :LINEPT1 :LINEPT2 STOP]
DRAWRESULT :POINT PARA VECTOR :LINEPT1 :LINEPT2
END

TO DRAWRESULT :POINT :VECT1
DRAWLINE :POINT SUM.POINT.VECT :POINT :VECT1
END
```

The purpose of each of the above procedures is summarized below, if
it has not been summarized on a previous page.

CP8 activates TURNOFFWRAP, SPLIT, DRAWCOORDS, and clears all text. It
    then moves on to DRAW.TRI.ALTS.
DRAW.TRI.ALTS runs all points through ADJUSTPOINT, then moves on to
    DRAWTRI and DRAW.ALTITUDES.
DRAW.ALTITUDES calls DRAW.PERP.OR.PARA to draw the line through each
    vertex that is perpendicular to the segment formed by the other
    two vertices.

Please note: This CP-page *does* stand alone, and all of the
procedures necessary to carry out its purpose are listed directly
after the example at the beginning. However, you may notice that
many of the above procedures have appeared previously. *If* it so
happens that you have already defined (typed in or loaded from disk)
all of the procedures from the CP2 page, the CP3 page, and the CP6
page since you started up LOGO, then LOGO already knows how to
perform many of the above procedures. If this is the case, the only
new procedures from this page that you need to type in are: CP8,
DRAW.TRI.ALTS, and DRAW.ALTITUDES.

Given a figure in the plane and a positive number R, magnify the figure by a factor of R

This page uses concepts from I15, I16, I19, and I20.

<u>Instructions</u>: To activate this set of procedures, type CP9 followed by the number of points in your figure (between 2 and 30), followed by your chosen magnification factor. Note that if you choose a very large magnification factor, the figure may be magnified right off the screen. Also note the magnification factor *can* be less than 1. After you type this, the program will prompt you for the coordinates of each point of the figure, as many times as the number of points you chose. When entering these points, use no brackets, or parentheses, or commas; simply type the two coordinates separated by a space, and hit ENTER (see below).

<u>Example</u>: To magnify by a factor of 3 the figure made up by the points (3,0), (0,2), (-2,0), and (0,-1), first type: CP9 4 3
     The program will then show:
                         TYPE IN COORDINATES OF FIGURE
                         WITHOUT BRACKETS, PARENTHESES OR COMMAS.
                         HIT ENTER AFTER TYPING EACH PAIR.
                         *** ENTER A COORDINATE PAIR:

You would then type each coordinate pair as shown, hitting the ENTER key after each one.
                         3 0
                         *** ENTER A COORDINATE PAIR:
                         0 2
                         *** ENTER A COORDINATE PAIR:
                         -2 0
                         *** ENTER A COORDINATE PAIR:
                         0 -1

When this is complete, the computer draws:

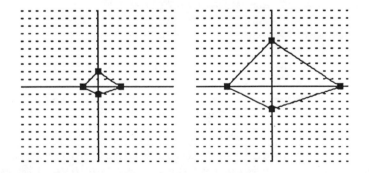

All of the procedures needed to carry out the above functions are listed below. They must be typed into the computer exactly as shown. Remember: any line that is indented from the margin is a continuation of the previous line.

```
TO CP9 :NUMOFPTS :MAGFACTOR
LOCAL "PTLIST
IF :NUMOFPTS < 2 [PR [2 POINTS MINIMUM!] STOP]
IF :NUMOFPTS > 3Ø [PR [3Ø POINTS MAXIMUM!] STOP]
TURNOFFWRAP
MAKE "PTLIST GET.POINTS :NUMOFPTS 6
SPLIT CLEARTEXT
DRAW.TWO.GRIDS
DRAW.FIGURE :NUMOFPTS PTLIST.XSHIFT :PTLIST -7Ø :NUMOFPTS 1 []
DRAW.FIGURE :NUMOFPTS PTLIST.XSHIFT (PTLIST.MAG :MAGFACTOR :PTLIST :NUMOFPTS 1
    []) 7Ø :NUMOFPTS 1 []
END


TO DRAW.TWO.GRIDS
HOME CS
PU SETPOS [7Ø Ø] DRAWGRID
PU SETPOS [-7Ø Ø] DRAWGRID
END


TO GET.POINTS :NUMOFPTS :ADJUSTFACTOR
LOCAL "POINTLIST
TEXTVIEW CLEARTEXT
PR [TYPE IN COORDINATES OF FIGURE]
PR [WITHOUT BRACKETS, PARENTHESES OR COMMAS.]
PR [HIT ENTER AFTER TYPING EACH PAIR.]
MAKE "POINTLIST []
REPEAT :NUMOFPTS [MAKE "POINTLIST PUT.PT.INTO.POINTLIST :ADJUSTFACTOR]
OP :POINTLIST
END


TO PUT.PT.INTO.POINTLIST :ADJUSTFACTOR
PR [*** ENTER A COORDINATE PAIR:]
OUTPUT LPUT SPROD :ADJUSTFACTOR READLIST :POINTLIST
END


TO DRAW.FIGURE :NUMOFPTS :POINTLIST
DRAW.ALL.PTS :POINTLIST :NUMOFPTS 1
CONNECT.THE.DOTS :POINTLIST :NUMOFPTS 1
END


TO DRAW.ALL.PTS :POINTLIST :NUMOFPTS :COUNTER
DRAWPT FETCHPT :COUNTER :POINTLIST
IF :COUNTER = :NUMOFPTS [STOP]
DRAW.ALL.PTS :POINTLIST :NUMOFPTS :COUNTER + 1
END


TO CONNECT.THE.DOTS :POINTLIST :NUMOFPTS :COUNTER
DRAWSEG FETCHPT :COUNTER :POINTLIST FETCHPT (:COUNTER + 1) :POINTLIST
IF (:COUNTER + 1) = :NUMOFPTS [DRAWSEG FETCHPT (:COUNTER + 1) :POINTLIST
    FETCHPT 1 :POINTLIST STOP]
CONNECT.THE.DOTS :POINTLIST :NUMOFPTS :COUNTER + 1
END


TO FETCHPT :PTNUM :LIST
OP ITEM :PTNUM :LIST
END


TO PT.XSHIFT :PAIR :XSHIFT
OP LIST :XSHIFT + FIRST :PAIR LAST :PAIR
END
```

```
TO PTLIST.XSHIFT :POINTLIST :XSHIFT :NUMOFPTS :COUNTER :NEWLIST
MAKE "NEWLIST LPUT PT.XSHIFT FIRST :POINTLIST :XSHIFT :NEWLIST
IF :COUNTER = :NUMOFPTS [OP :NEWLIST]
OP PTLIST.XSHIFT BUTFIRST :POINTLIST :XSHIFT :NUMOFPTS :COUNTER + 1 :NEWLIST
END


TO PT.MAG :PAIR :MAGFACTOR
OP SPROD :MAGFACTOR :PAIR
END


TO PTLIST.MAG :MAGFACTOR :POINTLIST :NUMOFPTS :COUNTER :NEWLIST
MAKE "NEWLIST LPUT PT.MAG FIRST :POINTLIST :MAGFACTOR :NEWLIST
IF :COUNTER = :NUMOFPTS [OP :NEWLIST]
OP PTLIST.MAG :MAGFACTOR BUTFIRST :POINTLIST :NUMOFPTS :COUNTER + 1 :NEWLIST
END


TO TEXTVIEW
TS
END


TO TURNOFFWRAP
HT WINDOW
END


TO SPLIT
SS
END


TO DRAWGRID
PD REPEAT 4 [FD 6Ø BK 6Ø RT 9Ø] PU
REPEAT 2 [RT 9Ø FD 6Ø] RT 9Ø
REPEAT 21 [REPEAT 21 [PD FD 1 PU FD 5] BK 126 RT 9Ø FD 6 LT 9Ø]
PU HOME
END


TO DRAWPT :PT
PU SETPOS :PT PD
MAKE "OLDHEAD HEADING
CENTERSQ 1
CENTERSQ 2
SETH :OLDHEAD
END


TO CENTERSQ :RADIUS
PU FD :RADIUS RT 9Ø
PD REPEAT 4 [FD :RADIUS RT 9Ø FD :RADIUS]
PU LT 9Ø BK :RADIUS
END


TO DRAWSEG :PT1 :PT2
PU SETPOS :PT1
PD SETPOS :PT2
END


TO SPROD :SCALAR :PAIR
OP LIST :SCALAR * (FIRST :PAIR) :SCALAR * (LAST :PAIR)
END
```

The purpose of each of the above procedures is summarized below, if it has not been summarized on a previous page.

CP9 first checks to make sure there are neither too many nor too few points. It then activates TURNOFFWRAP, stores the output list of GET.POINTS in :PTLIST, clears all text, and activates SPLIT plus DRAW.TWO.GRIDS. Then, it first draws the original unmagnified figure by activating DRAW.FIGURE with :PTLIST that has been moved by PTLIST.XSHIFT 70 units to the left. Next, it draws the magnified figure by activating DRAW.FIGURE with :PTLIST that has been run through PTLIST.MAG and moved 70 units to the right.

DRAW.TWO.GRIDS draws the right-hand and left-hand grids with DRAWGRID.

GET.POINTS activates TEXTVIEW, clears text, and prints point-entering instructions. It then makes :POINTLIST the output of repetitive calls of PUT.PT.INTO.POINTLIST, as many times as the entered number of points in the figure. It then outputs this list of points.

PUT.PT.INTO.POINTLIST asks for a coordinate pair to be entered, then puts the coordinate pair that is read from the user at the end of the list called :POINTLIST. This list thus becomes a list of lists, since each coordinate pair is actually a two-member list.

DRAW.FIGURE calls DRAW.ALL.PTS and CONNECT.THE.DOTS for the current list of points.

DRAW.ALL.PTS starts with a :COUNTER value of one. It uses FETCHPT to get the point out of the list of points whose position in the list is :COUNTER, draws that point, then repeats with a :COUNTER value that is one greater than before. This continues until the point that is drawn is the last one (:COUNTER = :NUMOFPTS).

CONNECT.THE.DOTS uses a counting system similar to DRAW.ALL.PTS to draw the segment from one point of the figure to the next. It draws a segment from the point whose position is :COUNTER to the point whose position is :COUNTER + 1. When the :COUNTER + 1 point is the last in the list, it draws a segment from this point to the first one, to close the figure, and stops.

FETCHPT outputs the point whose position in the list of points is :PTNUM.

PT.XSHIFT adds the entered :XSHIFT to the x-coordinate of the entered point.

PTLIST.XSHIFT again uses a counting mechanism to run each of the points in the list through PT.XSHIFT. When the procedure begins, :NEWLIST is empty and :POINTLIST is the list of all the points in the figure. First, the PT.XSHIFT of the first point in :POINTLIST is placed at the end of :NEWLIST. At the end of the procedure, it repeats, but the procedure uses BUTFIRST to re-activate itself with :POINTLIST containing all of the points the old :POINTLIST did, *except* for the first point. So the next time the first point in that list is shifted and put into :NEWLIST, it is actually the second point from the *original* :POINTLIST. This continues until each point from the original :POINTLIST has been altered and placed at the end of the growing :NEWLIST. At this time, :COUNTER = :NUMOFPTS and the procedure ceases.

PT.MAG magnifies a single point by giving an output that is the scalar product of the point and the magnification factor.

PTLIST.MAG uses a process similar to that found in PTLIST.XSHIFT to run each of the points in the list of points through PT.MAG.

TEXTVIEW creates a full-screen text display.

<u>Please note</u>: This CP-page *does* stand alone, and all of the
procedures necessary to carry out its purpose are listed directly
after the example at the beginning. However, you may notice that
many of the above procedures have appeared previously. *If* it so
happens that you have already defined (typed in or loaded from disk)
all of the procedures from the CP2 page since you started up LOGO,
then LOGO already knows how to perform some of the above procedures.
If this is the case, the only new procedures from this page that you
need to type in are: CP9, DRAW.TWO.GRIDS, GET.POINTS, PUT.PT.INTO.POINTLIST,
DRAW.FIGURE, DRAW.ALL.PTS, CONNECT.THE.DOTS, FETCHPT, PT.XSHIFT, PTLIST,XSHIFT,
PT.MAG, PTLIST.MAG, and TEXTVIEW.

Given a figure in the plane and two positive numbers R and S, magnify the figure by a factor of R in the horizontal direction and by a factor of S in the vertical direction

This page uses concepts from I32 and I32e.

Instructions: To activate this set of procedures, type CP10 followed by the number of points in your figure (between 2 and 30), followed by your chosen horizontal magnification factor R, followed by your chosen vertical magnification factor S. Note that if you choose very large magnification factors, the figure may be magnified right off the screen. Also note the magnification factors *can* be less than 1. After you type this, the program will prompt you for the coordinates of each point of the figure, as many times as the number of points you chose. When entering these points, use no brackets, or parentheses, or commas; simply type the two coordinates separated by a space, and hit ENTER (see below).

Example: To magnify, by a factor of 2 in the horizontal direction and by a factor of 0.5 in the vertical direction, the figure made up by the points (-5,2), (0,5), (5,2), (2,-4), and (-2,-4), first type:
CP10 5 2 0.5
   Then, the program will show:
                    TYPE IN COORDINATES OF FIGURE
                    WITHOUT BRACKETS, PARENTHESES OR COMMAS.
                    HIT ENTER AFTER TYPING EACH PAIR.
                    *** ENTER A COORDINATE PAIR:

You would then type each coordinate pair as shown, hitting the ENTER key after each one.
                    -5 2
                    *** ENTER A COORDINATE PAIR:
                    0 5
                    *** ENTER A COORDINATE PAIR:
                    5 2
                    *** ENTER A COORDINATE PAIR:
                    2 -4
                    *** ENTER A COORDINATE PAIR:
                    -2 -4

When this is complete, the computer draws:

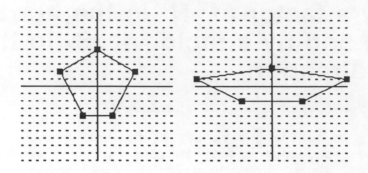

All of the procedures needed to carry out the above functions are
listed below. They must be typed into the computer exactly as
shown. Remember: any line that is indented from the margin is a
continuation of the previous line.

```
TO CP1Ø :NUMOFPTS :XMAG :YMAG
LOCAL "PTLIST
IF :NUMOFPTS < 2 [PR [2 POINTS MINIMUM!] STOP]
IF :NUMOFPTS > 3Ø [PR [3Ø POINTS MAXIMUM!] STOP]
TURNOFFWRAP
MAKE "PTLIST GET.POINTS :NUMOFPTS 6
SPLIT CLEARTEXT
DRAW.TWO.GRIDS
DRAW.FIGURE :NUMOFPTS PTLIST.XSHIFT :PTLIST -7Ø :NUMOFPTS 1 []
DRAW.FIGURE :NUMOFPTS PTLIST.XSHIFT (PTLIST.MAGXY :XMAG :YMAG :PTLIST :NUMOFPTS
    1 []) 7Ø :NUMOFPTS 1 []
END

TO DRAW.TWO.GRIDS
HOME CS
PU SETPOS [7Ø Ø] DRAWGRID
PU SETPOS [-7Ø Ø] DRAWGRID
END

TO DRAW.FIGURE :NUMOFPTS :POINTLIST
DRAW.ALL.PTS :POINTLIST :NUMOFPTS 1
CONNECT.THE.DOTS :POINTLIST :NUMOFPTS 1
END

TO DRAW.ALL.PTS :POINTLIST :NUMOFPTS :COUNTER
DRAWPT FETCHPT :COUNTER :POINTLIST
IF :COUNTER = :NUMOFPTS [STOP]
DRAW.ALL.PTS :POINTLIST :NUMOFPTS :COUNTER + 1
END

TO CONNECT.THE.DOTS :POINTLIST :NUMOFPTS :COUNTER
DRAWSEG FETCHPT :COUNTER :POINTLIST FETCHPT (:COUNTER + 1) :POINTLIST
IF (:COUNTER + 1) = :NUMOFPTS [DRAWSEG FETCHPT (:COUNTER + 1) :POINTLIST
    FETCHPT 1 :POINTLIST STOP]
CONNECT.THE.DOTS :POINTLIST :NUMOFPTS :COUNTER + 1
END
```

```
TO FETCHPT :PTNUM :LIST
OP ITEM :PTNUM :LIST
END

TO PT.XSHIFT :PAIR :XSHIFT
OP LIST :XSHIFT + FIRST :PAIR LAST :PAIR
END

TO PTLIST.XSHIFT :POINTLIST :XSHIFT :NUMOFPTS :COUNTER :NEWLIST
MAKE "NEWLIST LPUT PT.XSHIFT FIRST :POINTLIST :XSHIFT :NEWLIST
IF :COUNTER = :NUMOFPTS [OP :NEWLIST]
OP PTLIST.XSHIFT BUTFIRST :POINTLIST :XSHIFT :NUMOFPTS :COUNTER + 1 :NEWLIST
END

TO PT.MAGXY :PAIR :XMAG :YMAG
OP LIST (:XMAG * FIRST :PAIR) (:YMAG * LAST :PAIR)
END

TO PTLIST.MAGXY :XMAG :YMAG :POINTLIST :NUMOFPTS :COUNTER :NEWLIST
MAKE "NEWLIST LPUT PT.MAGXY FIRST :POINTLIST :XMAG :YMAG :NEWLIST
IF :COUNTER = :NUMOFPTS [OP :NEWLIST]
OP PTLIST.MAGXY :XMAG :YMAG BUTFIRST :POINTLIST :NUMOFPTS :COUNTER + 1 :NEWLIST
END

TO GET.POINTS :NUMOFPTS :ADJUSTFACTOR
LOCAL "POINTLIST
TEXTVIEW CLEARTEXT
PR [TYPE IN COORDINATES OF FIGURE]
PR [WITHOUT BRACKETS, PARENTHESES OR COMMAS.]
PR [HIT ENTER AFTER TYPING EACH PAIR.]
MAKE "POINTLIST []
REPEAT :NUMOFPTS [MAKE "POINTLIST PUT.PT.INTO.POINTLIST :ADJUSTFACTOR]
OP :POINTLIST
END

TO PUT.PT.INTO.POINTLIST :ADJUSTFACTOR
PR [*** ENTER A COORDINATE PAIR:]
OUTPUT LPUT SPROD :ADJUSTFACTOR READLIST :POINTLIST
END

TO TEXTVIEW
TS
END

TO TURNOFFWRAP
HT WINDOW
END

TO SPLIT
SS
END

TO DRAWGRID
PD REPEAT 4 [FD 60 BK 60 RT 90] PU
REPEAT 2 [RT 90 FD 60] RT 90
REPEAT 21 [REPEAT 21 [PD FD 1 PU FD 5] BK 126 RT 90 FD 6 LT 90]
PU HOME
END
```

```
TO DRAWPT :PT
PU SETPOS :PT PD
MAKE "OLDHEAD HEADING
CENTERSQ 1
CENTERSQ 2
SETH :OLDHEAD
END


TO CENTERSQ :RADIUS
PU FD :RADIUS RT 90
PD REPEAT 4 [FD :RADIUS RT 90 FD :RADIUS]
PU LT 90 BK :RADIUS
END


TO DRAWSEG :PT1 :PT2
PU SETPOS :PT1
PD SETPOS :PT2
END


TO SPROD :SCALAR :PAIR
OP LIST :SCALAR * (FIRST :PAIR) :SCALAR * (LAST :PAIR)
END
```

The purpose of each of the above procedures is summarized below, if
it has not been summarized on a previous page.

CP10 first checks to make sure there are neither too many nor too
    few points. It then activates TURNOFFWRAP, stores the output list
    of GET.POINTS in :PTLIST, clears all text, and activates SPLIT plus
    DRAW.TWO.GRIDS. Then, it first draws the original unmagnified
    figure by activating DRAW.FIGURE with :PTLIST that has been moved
    by PTLIST.XSHIFT 70 units to the left. Next, it draws the
    magnified figure by activating DRAW.FIGURE with :PTLIST that has
    been run through PTLIST.MAGXY and moved 70 units to the right.
PT.MAGXY multiplies the x-coordinate of the input point by R and the
    y-coordinate by S.
PTLIST.MAGXY runs each point in the list of points through PT.MAGXY in
    a manner similar to that used in PTLIST.XSHIFT (see description on
    CP9).

<u>Please note</u>: This CP-page *does* stand alone, and all of the
procedures necessary to carry out its purpose are listed directly
after the example at the beginning. However, you may notice that
many of the above procedures have appeared previously. *If* it so
happens that you have already defined (typed in or loaded from disk)
all of the procedures from the CP2 page and the CP9 page since you
started up LOGO, then LOGO already knows how to perform most of the
above procedures. If this is the case, the only new procedures from
this page that you need to type in are: CP10, PT.MAGXY, and
PTLIST.MAGXY.

Given the center and radius of a circle and two positive numbers R and S, magnify the circle by a factor of R in the horizontal direction and by a factor of S in the vertical direction

This page uses concepts from I32 and I32e.

<u>Instructions</u>: To activate this set of procedures, type CP11 followed by the coordinates of the center of the circle, followed by the radius of the circle, followed by the horizontal magnification factor R, followed by the vertical magnification factor S. Note that very large magnification factors may magnify the circle off the screen. Also note the magnification factors *can* be less than 1.

<u>Example</u>: To magnify the circle of center (1,-1) and radius 2 by a factor of 4 in the horizontal direction and by a factor of 2 in the vertical direction, type: CP11 [1 -1] 2 4 2
   The computer will draw:

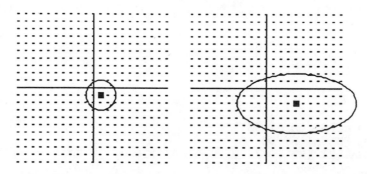

All of the procedures needed to carry out the above functions are listed below. They must be typed into the computer exactly as shown. Remember: any line that is indented from the margin is a continuation of the previous line.

```
TO CP11 :CNTR :RAD :XMAG :YMAG
TURNOFFWRAP
SPLIT CLEARTEXT
DRAW.TWO.GRIDS
DRAW.MAG.CIRCLE :CNTR :RAD :XMAG :YMAG
END

TO DRAW.MAG.CIRCLE :CNTR :RAD :XMAG :YMAG
MAKE "CNTR PT.XSHIFT ADJUSTPOINT :CNTR -7Ø
MAKE "RAD 6 * :RAD
DRAWPT :CNTR
DRAWPT MAGPOINT.SHIFT :CNTR :XMAG :YMAG
PU SETPOS :CNTR
SETH Ø FD :RAD
MAKE "OLDPT POS
REPEAT 12Ø [MAP.OUT.POINT :CNTR :RAD :XMAG :YMAG]
END
```

```
TO MAGPOINT.SHIFT :PAIR :XMAG :YMAG
OP PT.XSHIFT LIST :XMAG * FIRST PT.XSHIFT :PAIR 7Ø :YMAG * LAST PT.XSHIFT :PAIR
    7Ø 7Ø
END

TO PT.XSHIFT :PAIR :XSHIFT
OP LIST :XSHIFT + FIRST :PAIR LAST :PAIR
END

TO MAP.OUT.POINT :CNTR :RAD :XMAG :YMAG
PU SETPOS :CNTR
RT 3
FD :RAD
MAKE "NEWPT POS
DRAWSEG :OLDPT :NEWPT
DRAWSEG MAGPOINT.SHIFT :OLDPT :XMAG :YMAG MAGPOINT.SHIFT :NEWPT :XMAG :YMAG
MAKE "OLDPT :NEWPT
END

TO DRAW.TWO.GRIDS
HOME CS
PU SETPOS [7Ø Ø] DRAWGRID
PU SETPOS [-7Ø Ø] DRAWGRID
END

TO TURNOFFWRAP
HT WINDOW
END

TO SPLIT
SS
END

TO DRAWGRID
PD REPEAT 4 [FD 6Ø BK 6Ø RT 9Ø] PU
REPEAT 2 [RT 9Ø FD 6Ø] RT 9Ø
REPEAT 21 [REPEAT 21 [PD FD 1 PU FD 5] BK 126 RT 9Ø FD 6 LT 9Ø]
PU HOME
END

TO DRAWPT :PT
PU SETPOS :PT PD
MAKE "OLDHEAD HEADING
CENTERSQ 1
CENTERSQ 2
SETH :OLDHEAD
END

TO CENTERSQ :RADIUS
PU FD :RADIUS RT 9Ø
PD REPEAT 4 [FD :RADIUS RT 9Ø FD :RADIUS]
PU LT 9Ø BK :RADIUS
END

TO DRAWSEG :PT1 :PT2
PU SETPOS :PT1
PD SETPOS :PT2
END

TO ADJUSTPOINT :P
OP SPROD 6 :P
END
```

```
TO SPROD :SCALAR :PAIR
OP LIST :SCALAR * (FIRST :PAIR) :SCALAR * (LAST :PAIR)
END
```

The purpose of each of the above procedures is summarized below, if it has not been summarized on a previous page.

CP11 first activates TURNOFFWRAP, clears all text, and activates SPLIT
   plus DRAW.TWO.GRIDS. Then, it proceeds directly to DRAW.MAG.CIRCLE.
DRAW.MAG.CIRCLE first runs the center coordinates through ADJUSTPOINT
   while shifting them to the left-hand grid. It then multiplies
   the radius by 6 for the same adjustment. Then it draws the
   original and magnified centers. It then moves the turtle to the
   original center and prepares to draw. The current position is
   stored in :OLDPT, and MAP.OUT.POINT is repeated 120 times (each run
   covers three degrees, and 120x3˚=360˚).
MAGPOINT.SHIFT first shifts the figure on the left-hand grid to the
   center of the screen, where its coordinates can be accurately
   multiplied. It then multiplies the x and y coordinates by R and
   S, respectively, and shifts the resulting magnified point 70
   steps to the right to be on the right-hand grid.
MAP.OUT.POINT first puts the turtle on the original center, turns
   right three degrees, and goes out to the edge of the circle. Its
   current position then becomes :NEWPT. A segment is drawn from
   :OLDPT to :NEWPT, adding another piece to the circle. A segment
   is drawn between the MAGPOINT.SHIFTs of :OLDPT and :NEWPT, adding
   another piece to the ellipse. The current run's :NEWPT then
   becomes next run's :OLDPT. When the procedure repeats, a new
   :NEWPT is determined, and another segment drawn. So the circle
   and the ellipse are actually 120-sided polygons, but this doesn't
   matter. With the resolution of the computer screen being what it
   is, one can't tell the difference between a 120-gon and a circle.

Please note: This CP-page *does* stand alone, and all of the
procedures necessary to carry out its purpose are listed directly
after the example at the beginning. However, you may notice that
many of the above procedures have appeared previously. *If* it so
happens that you have already defined (typed in or loaded from disk)
all of the procedures from the CP2 page and the CP9 page since you
started up LOGO, then LOGO already knows how to perform most of the
above procedures. If this is the case, the only new procedures from
this page that you need to type in are: CP11, DRAW.MAG.CIRCLE,
MAGPOINT.SHIFT, and MAP.OUT.POINT.

TRANSLATIONS: Given a figure in the plane and two numbers a and b, show the motion m(x,y) = (x + a, y + b)

This page uses concepts from P6.

Instructions: To activate this set of procedures, type CP12 followed by the number of points in your figure (between 2 and 30), followed by your chosen numbers **a** and **b**. Note that if you choose a very large a or b, the figure may be moved right off the screen. After you type this, the program will prompt you for the coordinates of each point of the figure, as many times as the number of points you chose. When entering these points, use no brackets, or parentheses, or commas; simply type the two coordinates separated by a space, and hit ENTER (see below).

Example: To perform a translation, with a = −5 and b = 3, on the figure made up by the points (1,1), (3,1), (7,4), and (3,4), first type: CP12 4 −5 3
    The program will then show:
                    TYPE IN COORDINATES OF FIGURE
                    WITHOUT BRACKETS, PARENTHESES OR COMMAS.
                    HIT ENTER AFTER TYPING EACH PAIR.
                    *** ENTER A COORDINATE PAIR:

You would then type each coordinate pair as shown, hitting the ENTER key after each one.
                    1 1
                    *** ENTER A COORDINATE PAIR:
                    3 1
                    *** ENTER A COORDINATE PAIR:
                    7 4
                    *** ENTER A COORDINATE PAIR:
                    3 4

When this is complete, the computer draws:

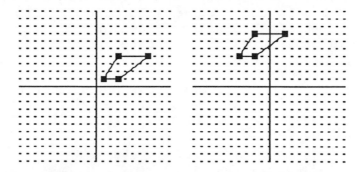

All of the procedures needed to carry out the above functions are listed below. They must be typed into the computer exactly as shown. Remember: any line that is indented from the margin is a continuation of the previous line.

```
TO CP12 :NUMOFPTS :XTRANS :YTRANS
LOCAL "PTLIST
IF :NUMOFPTS < 2 [PR [2 POINTS MINIMUM!] STOP]
IF :NUMOFPTS > 3Ø [PR [3Ø POINTS MAXIMUM!] STOP]
TURNOFFWRAP
MAKE "PTLIST GET.POINTS :NUMOFPTS 6
MAKE "XTRANS 6 * :XTRANS
MAKE "YTRANS 6 * :YTRANS
SPLIT CLEARTEXT
DRAW.TWO.GRIDS
DRAW.FIGURE :NUMOFPTS PTLIST.XSHIFT :PTLIST -7Ø :NUMOFPTS 1 []
DRAW.FIGURE :NUMOFPTS PTLIST.XSHIFT (PTLIST.TRANS :XTRANS :YTRANS :PTLIST
    :NUMOFPTS 1 []) 7Ø :NUMOFPTS 1 []
END


TO DRAW.TWO.GRIDS
HOME CS
PU SETPOS [7Ø Ø] DRAWGRID
PU SETPOS [-7Ø Ø] DRAWGRID
END


TO DRAW.FIGURE :NUMOFPTS :POINTLIST
DRAW.ALL.PTS :POINTLIST :NUMOFPTS 1
CONNECT.THE.DOTS :POINTLIST :NUMOFPTS 1
END


TO DRAW.ALL.PTS :POINTLIST :NUMOFPTS :COUNTER
DRAWPT FETCHPT :COUNTER :POINTLIST
IF :COUNTER = :NUMOFPTS [STOP]
DRAW.ALL.PTS :POINTLIST :NUMOFPTS :COUNTER + 1
END


TO CONNECT.THE.DOTS :POINTLIST :NUMOFPTS :COUNTER
DRAWSEG FETCHPT :COUNTER :POINTLIST FETCHPT (:COUNTER + 1) :POINTLIST
IF (:COUNTER + 1) = :NUMOFPTS [DRAWSEG FETCHPT (:COUNTER + 1) :POINTLIST
    FETCHPT 1 :POINTLIST STOP]
CONNECT.THE.DOTS :POINTLIST :NUMOFPTS :COUNTER + 1
END


TO FETCHPT :PTNUM :LIST
OP ITEM :PTNUM :LIST
END


TO PT.XSHIFT :PAIR :XSHIFT
OP LIST :XSHIFT + FIRST :PAIR LAST :PAIR
END


TO PTLIST.XSHIFT :POINTLIST :XSHIFT :NUMOFPTS :COUNTER :NEWLIST
MAKE "NEWLIST LPUT PT.XSHIFT FIRST :POINTLIST :XSHIFT :NEWLIST
IF :COUNTER = :NUMOFPTS [OP :NEWLIST]
OP PTLIST.XSHIFT BUTFIRST :POINTLIST :XSHIFT :NUMOFPTS :COUNTER + 1 :NEWLIST
END


TO PT.TRANS :PAIR :XTRANS :YTRANS
OP LIST (:XTRANS + FIRST :PAIR) (:YTRANS + LAST :PAIR)
END


TO PTLIST.TRANS :XTRANS :YTRANS :POINTLIST :NUMOFPTS :COUNTER :NEWLIST
MAKE "NEWLIST LPUT PT.TRANS FIRST :POINTLIST :XTRANS :YTRANS :NEWLIST
IF :COUNTER = :NUMOFPTS [OP :NEWLIST]
OP PTLIST.TRANS :XTRANS :YTRANS BUTFIRST :POINTLIST :NUMOFPTS :COUNTER + 1
    :NEWLIST
END
```

```
TO GET.POINTS :NUMOFPTS :ADJUSTFACTOR
LOCAL "POINTLIST
TEXTVIEW CLEARTEXT
PR [TYPE IN COORDINATES OF FIGURE]
PR [WITHOUT BRACKETS, PARENTHESES OR COMMAS.]
PR [HIT ENTER AFTER TYPING EACH PAIR.]
MAKE "POINTLIST []
REPEAT :NUMOFPTS [MAKE "POINTLIST PUT.PT.INTO.POINTLIST :ADJUSTFACTOR]
OP :POINTLIST
END


TO PUT.PT.INTO.POINTLIST :ADJUSTFACTOR
PR [*** ENTER A COORDINATE PAIR:]
OUTPUT LPUT SPROD :ADJUSTFACTOR READLIST :POINTLIST
END


TO TEXTVIEW
TS
END


TO TURNOFFWRAP
HT WINDOW
END


TO SPLIT
SS
END


TO DRAWGRID
PD REPEAT 4 [FD 6Ø BK 6Ø RT 9Ø] PU
REPEAT 2 [RT 9Ø FD 6Ø] RT 9Ø
REPEAT 21 [REPEAT 21 [PD FD 1 PU FD 5] BK 126 RT 9Ø FD 6 LT 9Ø]
PU HOME
END


TO DRAWPT :PT
PU SETPOS :PT PD
MAKE "OLDHEAD HEADING
CENTERSQ 1
CENTERSQ 2
SETH :OLDHEAD
END


TO CENTERSQ :RADIUS
PU FD :RADIUS RT 9Ø
PD REPEAT 4 [FD :RADIUS RT 9Ø FD :RADIUS]
PU LT 9Ø BK :RADIUS
END


TO DRAWSEG :PT1 :PT2
PU SETPOS :PT1
PD SETPOS :PT2
END


TO SPROD :SCALAR :PAIR
OP LIST :SCALAR * (FIRST :PAIR) :SCALAR * (LAST :PAIR)
END
```

The purpose of each of the above procedures is summarized below, if
it has not been summarized on a previous page.

CP12 first checks to make sure there are neither too many nor too
     few points. It then activates TURNOFFWRAP and stores the output
     list of GET.POINTS in :PTLIST. It multiplies a and b by 6 to
     fit the grid. It then clears all text, and activates SPLIT plus
     DRAW.TWO.GRIDS. It first draws the original unmodified figure by
     activating DRAW.FIGURE with :PTLIST that has been moved by
     PTLIST.XSHIFT 70 units to the left. Next, it draws the result of
     the translation by activating DRAW.FIGURE with :PTLIST that has
     been run through PTLIST.TRANS and moved 70 units to the right.
PT.TRANS performs the translation on a single point by adding a to
     the x-coordinate and adding b to the y-coordinate.
PTLIST.TRANS runs each point in the list of points through PT.TRANS in
     a manner similar to that used in PTLIST.XSHIFT (see description on
     CP9).

<u>Please note</u>: This CP-page *does* stand alone, and all of the
procedures necessary to carry out its purpose are listed directly
after the example at the beginning. However, you may notice that
many of the above procedures have appeared previously. *If* it so
happens that you have already defined (typed in or loaded from disk)
all of the procedures from the CP2 page and the CP9 page since you
started up LOGO, then LOGO already knows how to perform most of the
above procedures. If this is the case, the only new procedures from
this page that you need to type in are: CP12, PT.TRANS, and
PTLIST.TRANS.

ROTATIONS: Given a figure in the plane and two numbers c and s, so that $c^2 + s^2 = 1$, show the motion $m(x,y) = (cx - sy,\ \ sx + cy)$

This page uses concepts from P7.

<u>Instructions</u>: To activate this set of procedures, type CP13 followed by the number of points in your figure (between 2 and 30), followed by your chosen numbers **c** and **s**. Be sure to check your numbers beforehand with a calculator, so that $c^2 + s^2$ is exactly, or close enough to, one. After you type this, the program will calculate and display $c^2 + s^2$ for your choices of c and s ($c^2 + s^2$ between .95 and 1.05 is usually fine). Just hit any key, and the program proceeds. It then prompts you for the coordinates of each point of the figure, as many times as the number of points you chose. When entering these points, use no brackets, or parentheses, or commas; simply type the two coordinates separated by a space, and hit ENTER (see below).

<u>Example</u>: To perform a rotation, with c = -0.866 and b = -0.5, on the figure made up by the points (1,1), (3,1), (7,4), and (3,4), first type: CP13 4 -0.866 -0.5

    The program will then show something similar to:
```
                    C SQUARED PLUS S SQUARED EQUALS:
                    0.999956
                    PRESS ANY KEY TO CONTINUE
```

Then hit any key for the program to continue.

```
                    TYPE IN COORDINATES OF FIGURE
                    WITHOUT BRACKETS, PARENTHESES OR COMMAS.
                    HIT ENTER AFTER TYPING EACH PAIR.
                    *** ENTER A COORDINATE PAIR:
```

You would then type each coordinate pair as shown, hitting the ENTER key after each one.
```
                    1 1
                    *** ENTER A COORDINATE PAIR:
                    3 1
                    *** ENTER A COORDINATE PAIR:
                    7 4
                    *** ENTER A COORDINATE PAIR:
                    3 4
```

When this is complete, the computer draws:

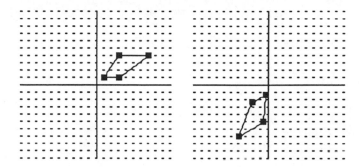

All of the procedures needed to carry out the above functions are listed below. They must be typed into the computer exactly as shown. Remember: any line that is indented from the margin is a continuation of the previous line.

```
TO CP13 :NUMOFPTS :C :S
VERIFY.ROTATION :C :S
LOCAL "PTLIST
IF :NUMOFPTS < 2 [PR [2 POINTS MINIMUM!] STOP]
IF :NUMOFPTS > 3Ø [PR [3Ø POINTS MAXIMUM!] STOP]
TURNOFFWRAP
MAKE "PTLIST GET.POINTS :NUMOFPTS 6
SPLIT CLEARTEXT
DRAW.TWO.GRIDS
DRAW.FIGURE :NUMOFPTS PTLIST.XSHIFT :PTLIST -7Ø :NUMOFPTS 1 []
DRAW.FIGURE :NUMOFPTS PTLIST.XSHIFT (PTLIST.ROT :C :S :PTLIST :NUMOFPTS 1 [])
    7Ø :NUMOFPTS 1 []
END

TO VERIFY.ROTATION :C :S
PR [C SQUARED PLUS S SQUARED EQUALS:]
PR (:C * :C) + (:S * :S)
WAITKEY
END

TO DRAW.TWO.GRIDS
HOME CS
PU SETPOS [7Ø Ø] DRAWGRID
PU SETPOS [-7Ø Ø] DRAWGRID
END

TO DRAW.FIGURE :NUMOFPTS :POINTLIST
DRAW.ALL.PTS :POINTLIST :NUMOFPTS 1
CONNECT.THE.DOTS :POINTLIST :NUMOFPTS 1
END

TO DRAW.ALL.PTS :POINTLIST :NUMOFPTS :COUNTER
DRAWPT FETCHPT :COUNTER :POINTLIST
IF :COUNTER = :NUMOFPTS [STOP]
DRAW.ALL.PTS :POINTLIST :NUMOFPTS :COUNTER + 1
END
```

```
TO CONNECT.THE.DOTS :POINTLIST :NUMOFPTS :COUNTER
DRAWSEG FETCHPT :COUNTER :POINTLIST FETCHPT (:COUNTER + 1) :POINTLIST
IF (:COUNTER + 1) = :NUMOFPTS [DRAWSEG FETCHPT (:COUNTER + 1) :POINTLIST
    FETCHPT 1 :POINTLIST STOP]
CONNECT.THE.DOTS :POINTLIST :NUMOFPTS :COUNTER + 1
END

TO FETCHPT :PTNUM :LIST
OP ITEM :PTNUM :LIST
END

TO PT.XSHIFT :PAIR :XSHIFT
OP LIST :XSHIFT + FIRST :PAIR LAST :PAIR
END

TO PTLIST.XSHIFT :POINTLIST :XSHIFT :NUMOFPTS :COUNTER :NEWLIST
MAKE "NEWLIST LPUT PT.XSHIFT FIRST :POINTLIST :XSHIFT :NEWLIST
IF :COUNTER = :NUMOFPTS [OP :NEWLIST]
OP PTLIST.XSHIFT BUTFIRST :POINTLIST :XSHIFT :NUMOFPTS :COUNTER + 1 :NEWLIST
END

TO PT.ROT :PAIR :C :S
OP LIST ((:C * FIRST :PAIR) - (:S * LAST :PAIR)) ((:S * FIRST :PAIR) + (:C *
    LAST :PAIR))
END

TO PTLIST.ROT :C :S :POINTLIST :NUMOFPTS :COUNTER :NEWLIST
MAKE "NEWLIST LPUT PT.ROT FIRST :POINTLIST :C :S :NEWLIST
IF :COUNTER = :NUMOFPTS [OP :NEWLIST]
OP PTLIST.ROT :C :S BUTFIRST :POINTLIST :NUMOFPTS :COUNTER + 1 :NEWLIST
END

TO GET.POINTS :NUMOFPTS :ADJUSTFACTOR
LOCAL "POINTLIST
TEXTVIEW CLEARTEXT
PR [TYPE IN COORDINATES OF FIGURE]
PR [WITHOUT BRACKETS, PARENTHESES OR COMMAS.]
PR [HIT ENTER AFTER TYPING EACH PAIR.]
MAKE "POINTLIST []
REPEAT :NUMOFPTS [MAKE "POINTLIST PUT.PT.INTO.POINTLIST :ADJUSTFACTOR]
OP :POINTLIST
END

TO PUT.PT.INTO.POINTLIST :ADJUSTFACTOR
PR [*** ENTER A COORDINATE PAIR:]
OUTPUT LPUT SPROD :ADJUSTFACTOR READLIST :POINTLIST
END

TO TEXTVIEW
TS
END

TO TURNOFFWRAP
HT WINDOW
END

TO SPLIT
SS
END
```

```
TO DRAWGRID
PD REPEAT 4 [FD 6Ø BK 6Ø RT 9Ø] PU
REPEAT 2 [RT 9Ø FD 6Ø] RT 9Ø
REPEAT 21 [REPEAT 21 [PD FD 1 PU FD 5] BK 126 RT 9Ø FD 6 LT 9Ø]
PU HOME
END

TO DRAWPT :PT
PU SETPOS :PT PD
MAKE "OLDHEAD HEADING
CENTERSQ 1
CENTERSQ 2
SETH :OLDHEAD
END

TO CENTERSQ :RADIUS
PU FD :RADIUS RT 9Ø
PD REPEAT 4 [FD :RADIUS RT 9Ø FD :RADIUS]
PU LT 9Ø BK :RADIUS
END

TO DRAWSEG :PT1 :PT2
PU SETPOS :PT1
PD SETPOS :PT2
END

TO SPROD :SCALAR :PAIR
OP LIST :SCALAR * (FIRST :PAIR) :SCALAR * (LAST :PAIR)
END

TO WAITKEY
PR [PRESS ANY KEY TO CONTINUE]
IGNORE RC
END

TO IGNORE :X
END
```

The purpose of each of the above procedures is summarized below, if
it has not been summarized on a previous page.

CP13 first activates VERIFY.ROTATION. It then checks to make sure
   there are neither too many nor too few points. It activates
   TURNOFFWRAP and stores the output list of GET.POINTS in :PTLIST. It
   then clears all text, and activates SPLIT plus DRAW.TWO.GRIDS. It
   first draws the original unmodified figure by activating
   DRAW.FIGURE with :PTLIST that has been moved by PTLIST.XSHIFT 70
   units to the left. Next, it draws the result of the rotation by
   activating DRAW.FIGURE with :PTLIST that has been run through
   PTLIST.ROT and moved 70 units to the right.
VERIFY.ROTATION prints out $c^2 + s^2$ and activates WAITKEY to wait for a
   key to be pressed before continuing.
PT.ROT performs the rotation on a single point by making an output
   with x-coordinate of cx - sy and y-coordinate of sx + cy.
PTLIST.ROT runs each point in the list of points through PT.ROT in a
   manner similar to that used in PTLIST.XSHIFT (see description on
   CP9).

Please note: This CP-page *does* stand alone, and all of the
procedures necessary to carry out its purpose are listed directly
after the example at the beginning. However, you may notice that
many of the above procedures have appeared previously. *If* it so
happens that you have already defined (typed in or loaded from disk)
all of the procedures from the CP2 page and the CP9 page since you
started up LOGO, then LOGO already knows how to perform most of the
above procedures. If this is the case, the only new procedures from
this page that you need to type in are: CP13, VERIFY.ROTATION, PT.ROT,
and PTLIST.ROT.

FLIPS: Given a figure in the plane, show the motion m(x,y) = (x , -y)

This page uses concepts from P8.

Instructions: To activate this set of procedures, type CP14
followed by the number of points in your figure (between 2 and 30).
After you type this, the program will prompt you for the coordinates
of each point of the figure, as many times as the number of points
you chose. When entering these points, use no brackets, or
parentheses, or commas; simply type the two coordinates separated by
a space, and hit ENTER (see below).

Example: To perform a flip on the figure made up by the points
(1,1), (3,1), (7,4), and (3,4), first type: CP14 4
    The program will then show:
                    TYPE IN COORDINATES OF FIGURE
                    WITHOUT BRACKETS, PARENTHESES OR COMMAS.
                    HIT ENTER AFTER TYPING EACH PAIR.
                    *** ENTER A COORDINATE PAIR:

You would then type each coordinate pair as shown, hitting the ENTER
key after each one.
                    1 1
                    *** ENTER A COORDINATE PAIR:
                    3 1
                    *** ENTER A COORDINATE PAIR:
                    7 4
                    *** ENTER A COORDINATE PAIR:
                    3 4

When this is complete, the computer draws:

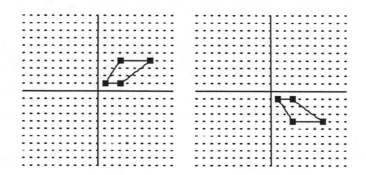

All of the procedures needed to carry out the above functions are
listed below. They must be typed into the computer exactly as
shown. Remember: any line that is indented from the margin is a
continuation of the previous line.

```
TO CP14 :NUMOFPTS
LOCAL "PTLIST
IF :NUMOFPTS < 2 [PR [2 POINTS MINIMUM!] STOP]
IF :NUMOFPTS > 3Ø [PR [3Ø POINTS MAXIMUM!] STOP]
TURNOFFWRAP
MAKE "PTLIST GET.POINTS :NUMOFPTS 6
SPLIT CLEARTEXT
DRAW.TWO.GRIDS
DRAW.FIGURE :NUMOFPTS PTLIST.XSHIFT :PTLIST -7Ø :NUMOFPTS 1 []
DRAW.FIGURE :NUMOFPTS PTLIST.XSHIFT (PTLIST.FLIP :PTLIST :NUMOFPTS 1 []) 7Ø
    :NUMOFPTS 1 []
END

TO DRAW.TWO.GRIDS
HOME CS
PU SETPOS [7Ø Ø] DRAWGRID
PU SETPOS [-7Ø Ø] DRAWGRID
END

TO DRAW.FIGURE :NUMOFPTS :POINTLIST
DRAW.ALL.PTS :POINTLIST :NUMOFPTS 1
CONNECT.THE.DOTS :POINTLIST :NUMOFPTS 1
END

TO DRAW.ALL.PTS :POINTLIST :NUMOFPTS :COUNTER
DRAWPT FETCHPT :COUNTER :POINTLIST
IF :COUNTER = :NUMOFPTS [STOP]
DRAW.ALL.PTS :POINTLIST :NUMOFPTS :COUNTER + 1
END

TO CONNECT.THE.DOTS :POINTLIST :NUMOFPTS :COUNTER
DRAWSEG FETCHPT :COUNTER :POINTLIST FETCHPT (:COUNTER + 1) :POINTLIST
IF (:COUNTER + 1) = :NUMOFPTS [DRAWSEG FETCHPT (:COUNTER + 1) :POINTLIST
    FETCHPT 1 :POINTLIST STOP]
CONNECT.THE.DOTS :POINTLIST :NUMOFPTS :COUNTER + 1
END

TO FETCHPT :PTNUM :LIST
OP ITEM :PTNUM :LIST
END

TO PT.XSHIFT :PAIR :XSHIFT
OP LIST :XSHIFT + FIRST :PAIR LAST :PAIR
END

TO PTLIST.XSHIFT :POINTLIST :XSHIFT :NUMOFPTS :COUNTER :NEWLIST
MAKE "NEWLIST LPUT PT.XSHIFT FIRST :POINTLIST :XSHIFT :NEWLIST
IF :COUNTER = :NUMOFPTS [OP :NEWLIST]
OP PTLIST.XSHIFT BUTFIRST :POINTLIST :XSHIFT :NUMOFPTS :COUNTER + 1 :NEWLIST
END

TO PT.FLIP :PAIR
OP LIST FIRST :PAIR (-1 * LAST :PAIR)
END

TO PTLIST.FLIP :POINTLIST :NUMOFPTS :COUNTER :NEWLIST
MAKE "NEWLIST LPUT PT.FLIP FIRST :POINTLIST :NEWLIST
IF :COUNTER = :NUMOFPTS [OP :NEWLIST]
OP PTLIST.FLIP BUTFIRST :POINTLIST :NUMOFPTS :COUNTER + 1 :NEWLIST
END
```

```
TO GET.POINTS :NUMOFPTS :ADJUSTFACTOR
LOCAL "POINTLIST
TEXTVIEW CLEARTEXT
PR [TYPE IN COORDINATES OF FIGURE]
PR [WITHOUT BRACKETS, PARENTHESES OR COMMAS.]
PR [HIT ENTER AFTER TYPING EACH PAIR.]
MAKE "POINTLIST []
REPEAT :NUMOFPTS [MAKE "POINTLIST PUT.PT.INTO.POINTLIST :ADJUSTFACTOR]
OP :POINTLIST
END


TO PUT.PT.INTO.POINTLIST :ADJUSTFACTOR
PR [*** ENTER A COORDINATE PAIR:]
OUTPUT LPUT SPROD :ADJUSTFACTOR READLIST :POINTLIST
END


TO TEXTVIEW
TS
END


TO TURNOFFWRAP
HT WINDOW
END


TO SPLIT
SS
END


TO DRAWGRID
PD REPEAT 4 [FD 6Ø BK 6Ø RT 9Ø] PU
REPEAT 2 [RT 9Ø FD 6Ø] RT 9Ø
REPEAT 21 [REPEAT 21 [PD FD 1 PU FD 5] BK 126 RT 9Ø FD 6 LT 9Ø]
PU HOME
END


TO DRAWPT :PT
PU SETPOS :PT PD
MAKE "OLDHEAD HEADING
CENTERSQ 1
CENTERSQ 2
SETH :OLDHEAD
END


TO CENTERSQ :RADIUS
PU FD :RADIUS RT 9Ø
PD REPEAT 4 [FD :RADIUS RT 9Ø FD :RADIUS]
PU LT 9Ø BK :RADIUS
END


TO DRAWSEG :PT1 :PT2
PU SETPOS :PT1
PD SETPOS :PT2
END


TO SPROD :SCALAR :PAIR
OP LIST :SCALAR * (FIRST :PAIR) :SCALAR * (LAST :PAIR)
END
```

The purpose of each of the above procedures is summarized below, if it has not been summarized on a previous page.

CP14 first checks to make sure there are neither too many nor too
few points. It activates TURNOFFWRAP and stores the output list
of GET.POINTS in :PTLIST. It then clears all text, and activates
SPLIT plus DRAW.TWO.GRIDS. It first draws the original unmodified
figure by activating DRAW.FIGURE with :PTLIST that has been moved
by PTLIST.XSHIFT 70 units to the left. Next, it draws the result
of the flip by activating DRAW.FIGURE with :PTLIST that has been
run through PTLIST.FLIP and moved 70 units to the right.
PT.FLIP performs the flip on a single point, and outputs a coordinate
pair whose y-coordinate has the opposite sign as that of the
input point.
PTLIST.FLIP runs each point in the list of points through PT.FLIP in a
manner similar to that used in PTLIST.XSHIFT (see description on
CP9).

Please note: This CP-page *does* stand alone, and all of the
procedures necessary to carry out its purpose are listed directly
after the example at the beginning. However, you may notice that
many of the above procedures have appeared previously. *If* it so
happens that you have already defined (typed in or loaded from disk)
all of the procedures from the CP2 page and the CP9 page since you
started up LOGO, then LOGO already knows how to perform most of the
above procedures. If this is the case, the only new procedures from
this page that you need to type in are: CP14, PT.FLIP, and PTLIST.FLIP.

Composing a set of two motions

This page uses concepts from P6, P7, P8, and P10.

<u>Instructions</u>: To activate this set of procedures, type CP15 followed by the number of points in your figure (between 2 and 30). After you type this, the program will prompt you for the coordinates of each point of the figure, as many times as the number of points you chose. When entering these points, use no brackets, or parentheses, or commas; simply type the two coordinates separated by a space, and hit ENTER (see below). After all points are entered, the program will prompt you to enter the two motions that make up the set. Translations are entered as the letter T, followed by the number a, followed by the number b. Rotations are entered as the letter R, followed by the number c, followed by the number s. Flips are entered as the letter F alone (see below). Whenever you enter a rotation, the program will print the $c^2 + s^2$ for the c and s that you entered. Press any key to continue the entering process.

<u>Example</u>: To perform a rotation (with c = 0.866 and s = -0.5) followed by a translation (with a = -5 and b = 10) on the figure made up by (5,4), (10,5), (10,12), and (5,8), first type: CP15 4
    The program will then show:

```
TYPE IN COORDINATES OF FIGURE
WITHOUT BRACKETS, PARENTHESES OR COMMAS.
HIT ENTER AFTER TYPING EACH PAIR.
*** ENTER A COORDINATE PAIR:
```

You would then type each coordinate pair as shown, hitting the ENTER key after each one.

```
5 4
*** ENTER A COORDINATE PAIR:
10 5
*** ENTER A COORDINATE PAIR:
10 12
*** ENTER A COORDINATE PAIR:
5 8
```

Note that figures can be seen best when the coordinates are a little larger than usual. These grids stretch to 15, not 10, units.
When this is complete, the computer asks for the motions:

```
*********************
EXAMPLES OF HOW TO ENTER MOTIONS:
TO TRANSLATE WITH A = 4 AND B = -2, TYPE: T 4 -2
TO ROTATE WITH C = Ø.866 AND S = Ø.5, TYPE: R .866 .5
TO FLIP VERTICALLY, TYPE: F
*** TYPE FIRST MOTION OF THIS SET AND HIT ENTER:
R 0.866 -0.5
```

Since this is a rotation, the program shows $c^2 + s^2$:

```
C SQUARED PLUS S SQUARED EQUALS:
0.999956
PRESS ANY KEY TO CONTINUE

*** TYPE SECOND MOTION OF THIS SET AND HIT ENTER
T -5 10
```

When this is complete, the computer draws:

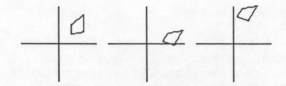

The left-hand grid shows the original figure, the center grid shows
the result of applying the first motion to the figure in the left-
hand grid, and the right-hand grid shows the result of applying the
second motion to the figure in the center grid.

All of the procedures needed to carry out the above functions are
listed below. They must be typed into the computer exactly as
shown. Remember: any line that is indented from the margin is a
continuation of the previous line.

```
TO CP15 :NUMOFPTS
IF :NUMOFPTS < 2 [PR [2 POINTS MINIMUM!] STOP]
IF :NUMOFPTS > 3Ø [PR [3Ø POINTS MAXIMUM!] STOP]
TEXTVIEW CLEARTEXT
MAKE "PTLIST GET.POINTS :NUMOFPTS 2
MAKE "TWOMOTIONS GET.MOTIONS
TURNOFFWRAP
MAKE "PTLIST DRAW.TWO.MOTIONS :PTLIST :TWOMOTIONS :NUMOFPTS
END


TO DRAW.TWO.MOTIONS :PTLIST :TWOMOTIONS :NUMOFPTS
HOME CS SPLIT CLEARTEXT
DRAW.MINI.AXES
CONNECT.THE.DOTS PTLIST.XSHIFT :PTLIST -7Ø :NUMOFPTS 1 [] :NUMOFPTS 1
MAKE "PTLIST CARRY.OUT.MOTION :PTLIST FIRST :TWOMOTIONS :NUMOFPTS
CONNECT.THE.DOTS :PTLIST :NUMOFPTS 1
MAKE "PTLIST CARRY.OUT.MOTION :PTLIST LAST :TWOMOTIONS :NUMOFPTS
CONNECT.THE.DOTS PTLIST.XSHIFT :PTLIST 7Ø :NUMOFPTS 1 [] :NUMOFPTS 1
OP :PTLIST
END


TO GET.MOTIONS
LOCAL "MOTIONSET
MAKE "MOTIONSET []
TEXTVIEW CLEARTEXT
PR [*********************] PR [ ]
PR [EXAMPLES OF HOW TO ENTER MOTIONS:]
PR [TO TRANSLATE WITH A = 4 AND B = -2, TYPE: T 4 -2]
PR [TO ROTATE WITH C = Ø.866 AND S = Ø.5, TYPE: R .866 .5]
PR [TO FLIP VERTICALLY, TYPE: F] PR [ ]
PR [*** TYPE FIRST MOTION OF THIS SET AND HIT ENTER:]
MAKE "MOTIONSET LPUT READLIST :MOTIONSET
IF (FIRST (FIRST :MOTIONSET)) = "R [VERIFY.ROTATION ITEM 2 (FIRST :MOTIONSET)
    ITEM 3 (FIRST :MOTIONSET)]
PR [*** TYPE SECOND MOTION OF THIS SET AND HIT ENTER:]
MAKE "MOTIONSET LPUT READLIST :MOTIONSET
IF (FIRST (LAST :MOTIONSET)) = "R [VERIFY.ROTATION ITEM 2 (LAST :MOTIONSET)
    ITEM 3 (LAST :MOTIONSET)]
SPLIT CLEARTEXT
OP :MOTIONSET
END
```

```
TO DRAW.MINI.AXES
PU SETPOS [-7Ø Ø] PD REPEAT 4 [FD 3Ø BK 3Ø RT 9Ø]
PU SETPOS [Ø Ø] PD REPEAT 4 [FD 3Ø BK 3Ø RT 9Ø]
PU SETPOS [7Ø Ø] PD REPEAT 4 [FD 3Ø BK 3Ø RT 9Ø]
END


TO CARRY.OUT.MOTION :POINTLIST :MOTION :NUMOFPTS
LOCAL "MOVEDLIST
IF (FIRST :MOTION) = "T [MAKE "MOVEDLIST PTLIST.TRANS (2 * ITEM 2 :MOTION) (2 *
    ITEM 3 :MOTION) :POINTLIST :NUMOFPTS 1 []]
IF (FIRST :MOTION) = "R [MAKE "MOVEDLIST PTLIST.ROT ITEM 2 :MOTION ITEM 3
    :MOTION :POINTLIST :NUMOFPTS 1 []]
IF (FIRST :MOTION) = "F [MAKE "MOVEDLIST PTLIST.FLIP :POINTLIST :NUMOFPTS 1 []]
OP :MOVEDLIST
END


TO GET.POINTS :NUMOFPTS :ADJUSTFACTOR
LOCAL "POINTLIST
TEXTVIEW CLEARTEXT
PR [TYPE IN COORDINATES OF FIGURE]
PR [WITHOUT BRACKETS, PARENTHESES OR COMMAS.]
PR [HIT ENTER AFTER TYPING EACH PAIR.]
MAKE "POINTLIST []
REPEAT :NUMOFPTS [MAKE "POINTLIST PUT.PT.INTO.POINTLIST :ADJUSTFACTOR]
OP :POINTLIST
END


TO PUT.PT.INTO.POINTLIST :ADJUSTFACTOR
PR [*** ENTER A COORDINATE PAIR:]
OUTPUT LPUT SPROD :ADJUSTFACTOR READLIST :POINTLIST
END


TO CONNECT.THE.DOTS :POINTLIST :NUMOFPTS :COUNTER
DRAWSEG FETCHPT :COUNTER :POINTLIST FETCHPT (:COUNTER + 1) :POINTLIST
IF (:COUNTER + 1) = :NUMOFPTS [DRAWSEG FETCHPT (:COUNTER + 1) :POINTLIST
    FETCHPT 1 :POINTLIST STOP]
CONNECT.THE.DOTS :POINTLIST :NUMOFPTS :COUNTER + 1
END


TO FETCHPT :PTNUM :LIST
OP ITEM :PTNUM :LIST
END


TO PT.XSHIFT :PAIR :XSHIFT
OP LIST :XSHIFT + FIRST :PAIR LAST :PAIR
END


TO PTLIST.XSHIFT :POINTLIST :XSHIFT :NUMOFPTS :COUNTER :NEWLIST
MAKE "NEWLIST LPUT PT.XSHIFT FIRST :POINTLIST :XSHIFT :NEWLIST
IF :COUNTER = :NUMOFPTS [OP :NEWLIST]
OP PTLIST.XSHIFT BUTFIRST :POINTLIST :XSHIFT :NUMOFPTS :COUNTER + 1 :NEWLIST
END


TO PT.TRANS :PAIR :XTRANS :YTRANS
OP LIST (:XTRANS + FIRST :PAIR) (:YTRANS + LAST :PAIR)
END


TO PTLIST.TRANS :XTRANS :YTRANS :POINTLIST :NUMOFPTS :COUNTER :NEWLIST
MAKE "NEWLIST LPUT PT.TRANS FIRST :POINTLIST :XTRANS :YTRANS :NEWLIST
IF :COUNTER = :NUMOFPTS [OP :NEWLIST]
OP PTLIST.TRANS :XTRANS :YTRANS BUTFIRST :POINTLIST :NUMOFPTS :COUNTER + 1
    :NEWLIST
END
```

```
TO PT.ROT :PAIR :C :S
OP LIST ((:C * FIRST :PAIR) - (:S * LAST :PAIR)) ((:S * FIRST :PAIR) + (:C *
    LAST :PAIR))
END

TO PTLIST.ROT :C :S :POINTLIST :NUMOFPTS :COUNTER :NEWLIST
MAKE "NEWLIST LPUT PT.ROT FIRST :POINTLIST :C :S :NEWLIST
IF :COUNTER = :NUMOFPTS [OP :NEWLIST]
OP PTLIST.ROT :C :S BUTFIRST :POINTLIST :NUMOFPTS :COUNTER + 1 :NEWLIST
END

TO PT.FLIP :PAIR
OP LIST FIRST :PAIR (-1 * LAST :PAIR)
END

TO PTLIST.FLIP :POINTLIST :NUMOFPTS :COUNTER :NEWLIST
MAKE "NEWLIST LPUT PT.FLIP FIRST :POINTLIST :NEWLIST
IF :COUNTER = :NUMOFPTS [OP :NEWLIST]
OP PTLIST.FLIP BUTFIRST :POINTLIST :NUMOFPTS :COUNTER + 1 :NEWLIST
END

TO VERIFY.ROTATION :C :S
PR [C SQUARED PLUS S SQUARED EQUALS:]
PR (:C * :C) + (:S * :S)
WAITKEY
END

TO WAITKEY
PR [PRESS ANY KEY TO CONTINUE]
IGNORE RC
END

TO IGNORE :X
END

TO TEXTVIEW
TS
END

TO TURNOFFWRAP
HT WINDOW
END

TO SPLIT
SS
END

TO DRAWSEG :PT1 :PT2
PU SETPOS :PT1
PD SETPOS :PT2
END

TO SPROD :SCALAR :PAIR
OP LIST :SCALAR * (FIRST :PAIR) :SCALAR * (LAST :PAIR)
END
```

The purpose of each of the above procedures is summarized below, if it has not been summarized on a previous page.

CP15 first checks to make sure there are neither too many nor too
   few points. It activates TEXTVIEW, clears all text, stores the
   output list of GET.POINTS in :PTLIST, and stores the output list of
   GET.MOTIONS in :TWOMOTIONS. It activates TURNOFFWRAP. It then
   stores the output list of DRAW.TWO.MOTIONS in :PTLIST.
DRAW.TWO.MOTIONS first clears the graphics and all text, and activates
   SPLIT and DRAW.MINI.AXES. The first call of CONNECT.THE.DOTS draws
   the original figure, straight from the entered pointlist, x-
   shifted to the left-hand grid. This point list is then run
   through CARRY.OUT.MOTION with the first entered motion as the
   motion, modifying the list. The new list is then drawn on the
   center grid with CONNECT.THE.DOTS. The list is again run through
   CARRY.OUT.MOTION, this time with the second entered motion. The
   resulting figure, which has undergone both entered motions, is x-
   shifted to the right-hand grid and drawn with CONNECT.THE.DOTS.
   The new point list is then output back to CP15.
GET.MOTIONS empties the :MOTIONSET variable, activates TEXTVIEW, clears
   the text, and prints instructions. The first motion is read from
   the keyboard, and placed at the end of :MOTIONSET. If it is a
   rotation, VERIFY.ROTATION is activated. The second motion is read
   from the keyboard, and placed at the end of :MOTIONSET. Again,
   VERIFY.ROTATION is activated if needed. After activating SPLIT and
   clearing text, the list of two motions is output back to CP15.
   Note that since each entered motion is a list itself, :MOTIONSET
   is a list of lists.
DRAW.MINI.AXES draws the three scaled-down (x,y)-axes.
CARRY.OUT.MOTION chooses how to modify :POINTLIST depending on the first
   letter of the :MOTION with which it is currently operating. If
   the first letter of the motion is T, the motion is a translation
   and the :POINTLIST is modified by running it through PTLIST.TRANS.
   If the letter is R, the list is run through PTLIST.ROT, and if it
   is F, the list is put through PTLIST.FLIP. Finally, the list that
   has been appropriately modified is output back to DRAW.TWO.MOTIONS.

<u>Please note</u>: This CP-page *does* stand alone, and all of the
procedures necessary to carry out its purpose are listed directly
after the example at the beginning. However, you may notice that
many of the above procedures have appeared previously. *If* it so
happens that you have already defined (typed in or loaded from disk)
all of the procedures from the CP2 page, the CP9 page, the CP12
page, the CP13 page, and the CP14 page since you started up LOGO,
then LOGO already knows how to perform most of the above procedures.
If this is the case, the only new procedures from this page that you
need to type in are: CP15, DRAW.TWO.MOTIONS, GET.MOTIONS, DRAW.MINI.AXES,
and CARRY.OUT.MOTION.

Composing a series of motions

This page uses concepts from P6, P7, P8, and P10.

<u>Instructions</u>: To activate this set of procedures, type CP16 followed by the number of points in your figure (between 2 and 30). After you type this, the program will prompt you for the coordinates of each point of the figure, as many times as the number of points you chose. When entering these points, use no brackets, or parentheses, or commas; simply type the two coordinates separated by a space, and hit ENTER (see below). After all points are entered, the program will prompt you to enter two motions that make up the first set. Translations are entered as the letter T, followed by the number a, followed by the number b. Rotations are entered as the letter R, followed by the number c, followed by the number s. Flips are entered as the letter F alone (see below). Whenever you enter a rotation, the program will print the $c^2 + s^2$ for the c and s that you entered. Press any key to continue the entering process. Once the two motions you entered have been drawn, the program will ask whether or not you want to continue. If you hit the S key, the program ends. If you hit M instead, the program will ask for two more motions just as before. This time, however, the motions will be applied to the resulting figure from the last set of two motions that you did. In other words, the figure in the right-hand grid during your last set of motions will be the figure in the left-hand grid during this set, and the two new motions will apply to *it*. In this way you can start with a figure, and essentially apply as many motions to it as you want. Doing this might help you understand Theorem 12 from P10, which says that no matter what type of motion you want to produce, it can be made from combinations of translations, rotations, and vertical flips. For instance, you might try to find a way to use a vertical flip in combination with a rotation in order to eventually produce a horizontal flip of the original figure, etc.

<u>Example</u>: Since this program operates in a manner similar to CP15, the whole example will not be repeated. The difference from CP15 is that after drawing each motion set, the program asks you to:

> HIT "M" TO CARRY OUT ANOTHER SET OF TWO MOTIONS
> ON THE RESULTING FIGURE. HIT "S" TO STOP.

Hitting M will continue the series of motions for another set.

All of the procedures needed to carry out the above functions are listed below. They must be typed into the computer exactly as shown. Remember: any line that is indented from the margin is a continuation of the previous line.

```
TO CP16 :NUMOFPTS
IF :NUMOFPTS < 2 [PR [2 POINTS MINIMUM!] STOP]
IF :NUMOFPTS > 3Ø [PR [3Ø POINTS MAXIMUM!] STOP]
TEXTVIEW CLEARTEXT
MAKE "PTLIST GET.POINTS :NUMOFPTS 2
TURNOFFWRAP SPLIT
DRAW.MANY.MOTIONS :PTLIST :NUMOFPTS
TEXTVIEW CLEARTEXT
END
```

```
TO DRAW.MANY.MOTIONS :PTLIST :NUMOFPTS
MAKE "TWOMOTIONS GET.MOTIONS
MAKE "PTLIST DRAW.TWO.MOTIONS :PTLIST :TWOMOTIONS :NUMOFPTS
CLEARTEXT
PR [HIT "M" TO CARRY OUT ANOTHER SET OF TWO MOTIONS]
PR [ON THE RESULTING FIGURE.  HIT "S" TO STOP.]
IF RC = "M [DRAW.MANY.MOTIONS :PTLIST :NUMOFPTS]
END


TO DRAW.TWO.MOTIONS :PTLIST :TWOMOTIONS :NUMOFPTS
HOME CS SPLIT CLEARTEXT
DRAW.MINI.AXES
CONNECT.THE.DOTS (PTLIST.XSHIFT :PTLIST -7Ø :NUMOFPTS 1 []) :NUMOFPTS 1
MAKE "PTLIST CARRY.OUT.MOTION :PTLIST FIRST :TWOMOTIONS :NUMOFPTS
CONNECT.THE.DOTS :PTLIST :NUMOFPTS 1
MAKE "PTLIST CARRY.OUT.MOTION :PTLIST LAST :TWOMOTIONS :NUMOFPTS
CONNECT.THE.DOTS (PTLIST.XSHIFT :PTLIST 7Ø :NUMOFPTS 1 []) :NUMOFPTS 1
OP :PTLIST
END


TO DRAW.MINI.AXES
PU SETPOS [-7Ø Ø] PD REPEAT 4 [FD 3Ø BK 3Ø RT 9Ø]
PU SETPOS [Ø Ø] PD REPEAT 4 [FD 3Ø BK 3Ø RT 9Ø]
PU SETPOS [7Ø Ø] PD REPEAT 4 [FD 3Ø BK 3Ø RT 9Ø]
END


TO CARRY.OUT.MOTION :POINTLIST :MOTION :NUMOFPTS
LOCAL "MOVEDLIST
IF (FIRST :MOTION) = "T [MAKE "MOVEDLIST PTLIST.TRANS (2 * ITEM 2 :MOTION) (2 *
    ITEM 3 :MOTION) :POINTLIST :NUMOFPTS 1 []]
IF (FIRST :MOTION) = "R [MAKE "MOVEDLIST PTLIST.ROT ITEM 2 :MOTION ITEM 3
    :MOTION :POINTLIST :NUMOFPTS 1 []]
IF (FIRST :MOTION) = "F [MAKE "MOVEDLIST PTLIST.FLIP :POINTLIST :NUMOFPTS 1 []]
OP :MOVEDLIST
END


TO GET.POINTS :NUMOFPTS :ADJUSTFACTOR
LOCAL "POINTLIST
TEXTVIEW CLEARTEXT
PR [TYPE IN COORDINATES OF FIGURE]
PR [WITHOUT BRACKETS, PARENTHESES OR COMMAS.]
PR [HIT ENTER AFTER TYPING EACH PAIR.]
MAKE "POINTLIST []
REPEAT :NUMOFPTS [MAKE "POINTLIST PUT.PT.INTO.POINTLIST :ADJUSTFACTOR]
OP :POINTLIST
END


TO PUT.PT.INTO.POINTLIST :ADJUSTFACTOR
PR [*** ENTER A COORDINATE PAIR:]
OUTPUT LPUT SPROD :ADJUSTFACTOR READLIST :POINTLIST
END
```

```
TO GET.MOTIONS
LOCAL "MOTIONSET
MAKE "MOTIONSET []
TEXTVIEW CLEARTEXT
PR [********************] PR [ ]
PR [EXAMPLES OF HOW TO ENTER MOTIONS:]
PR [TO TRANSLATE WITH A = 4 AND B = -2, TYPE: T 4 -2]
PR [TO ROTATE WITH C = Ø.866 AND S = Ø.5, TYPE: R .866 .5]
PR [TO FLIP VERTICALLY, TYPE: F] PR [ ]
PR [*** TYPE FIRST MOTION OF THIS SET AND HIT ENTER:]
MAKE "MOTIONSET LPUT READLIST :MOTIONSET
IF (FIRST (FIRST :MOTIONSET)) = "R [VERIFY.ROTATION ITEM 2 (FIRST :MOTIONSET)
    ITEM 3 (FIRST :MOTIONSET)]
PR [*** TYPE SECOND MOTION OF THIS SET AND HIT ENTER:]
MAKE "MOTIONSET LPUT READLIST :MOTIONSET
IF (FIRST (LAST :MOTIONSET)) = "R [VERIFY.ROTATION ITEM 2 (LAST :MOTIONSET)
    ITEM 3 (LAST :MOTIONSET)]
SPLIT CLEARTEXT
OP :MOTIONSET
END


TO CONNECT.THE.DOTS :POINTLIST :NUMOFPTS :COUNTER
DRAWSEG FETCHPT :COUNTER :POINTLIST FETCHPT (:COUNTER + 1) :POINTLIST
IF (:COUNTER + 1) = :NUMOFPTS [DRAWSEG FETCHPT (:COUNTER + 1) :POINTLIST
    FETCHPT 1 :POINTLIST STOP]
CONNECT.THE.DOTS :POINTLIST :NUMOFPTS :COUNTER + 1
END


TO FETCHPT :PTNUM :LIST
OP ITEM :PTNUM :LIST
END


TO PT.XSHIFT :PAIR :XSHIFT
OP LIST :XSHIFT + FIRST :PAIR LAST :PAIR
END


TO PTLIST.XSHIFT :POINTLIST :XSHIFT :NUMOFPTS :COUNTER :NEWLIST
MAKE "NEWLIST LPUT PT.XSHIFT FIRST :POINTLIST :XSHIFT :NEWLIST
IF :COUNTER = :NUMOFPTS [OP :NEWLIST]
OP PTLIST.XSHIFT BUTFIRST :POINTLIST :XSHIFT :NUMOFPTS :COUNTER + 1 :NEWLIST
END


TO PT.TRANS :PAIR :XTRANS :YTRANS
OP LIST (:XTRANS + FIRST :PAIR) (:YTRANS + LAST :PAIR)
END


TO PTLIST.TRANS :XTRANS :YTRANS :POINTLIST :NUMOFPTS :COUNTER :NEWLIST
MAKE "NEWLIST LPUT PT.TRANS FIRST :POINTLIST :XTRANS :YTRANS :NEWLIST
IF :COUNTER = :NUMOFPTS [OP :NEWLIST]
OP PTLIST.TRANS :XTRANS :YTRANS BUTFIRST :POINTLIST :NUMOFPTS :COUNTER + 1
    :NEWLIST
END


TO PT.ROT :PAIR :C :S
OP LIST ((:C * FIRST :PAIR) - (:S * LAST :PAIR)) ((:S * FIRST :PAIR) + (:C *
    LAST :PAIR))
END


TO PTLIST.ROT :C :S :POINTLIST :NUMOFPTS :COUNTER :NEWLIST
MAKE "NEWLIST LPUT PT.ROT FIRST :POINTLIST :C :S :NEWLIST
IF :COUNTER = :NUMOFPTS [OP :NEWLIST]
OP PTLIST.ROT :C :S BUTFIRST :POINTLIST :NUMOFPTS :COUNTER + 1 :NEWLIST
END
```

```
TO PT.FLIP :PAIR
OP LIST FIRST :PAIR (-1 * LAST :PAIR)
END

TO PTLIST.FLIP :POINTLIST :NUMOFPTS :COUNTER :NEWLIST
MAKE "NEWLIST LPUT PT.FLIP FIRST :POINTLIST :NEWLIST
IF :COUNTER = :NUMOFPTS [OP :NEWLIST]
OP PTLIST.FLIP BUTFIRST :POINTLIST :NUMOFPTS :COUNTER + 1 :NEWLIST
END

TO VERIFY.ROTATION :C :S
PR [C SQUARED PLUS S SQUARED EQUALS:]
PR (:C * :C) + (:S * :S)
WAITKEY
END

TO WAITKEY
PR [PRESS ANY KEY TO CONTINUE]
IGNORE RC
END

TO IGNORE :X
END

TO TEXTVIEW
TS
END

TO TURNOFFWRAP
HT WINDOW
END

TO SPLIT
SS
END

TO DRAWSEG :PT1 :PT2
PU SETPOS :PT1
PD SETPOS :PT2
END

TO SPROD :SCALAR :PAIR
OP LIST :SCALAR * (FIRST :PAIR) :SCALAR * (LAST :PAIR)
END
```

The purpose of each of the above procedures is summarized below, if
it has not been summarized on a previous page.

CP16 first checks to make sure there are neither too many nor too
  few points. It activates TEXTVIEW, clears all text, and stores
  the output list of GET.POINTS in :PTLIST. It activates TURNOFFWRAP
  and SPLIT. It then activates DRAW.MANY.MOTIONS, and once this cycle
  is ended by the user, clears the text and stops.
DRAW.MANY.MOTIONS first stores the output of GET.MOTIONS in :TWOMOTIONS.
  These are the two motions that will be carried out this time. It
  then stores the output of DRAW.TWO.MOTIONS in :PTLIST. This causes
  the :PTLIST to contain the points that resulted from the last two
  motions performed. After printing instructions, the procedure

repeats itself, thereby getting and carrying out two more
motions, but only if the key that was hit was M.

Please note: This CP-page *does* stand alone, and all of the
procedures necessary to carry out its purpose are listed directly
after the example at the beginning. However, you may notice that
many of the above procedures have appeared previously. *If* it so
happens that you have already defined (typed in or loaded from disk)
all of the procedures from the CP2 page, the CP9 page, the CP12
page, the CP13 page, the CP14 page, and the CP15 page since you
started up LOGO, then LOGO already knows how to perform most of the
above procedures. If this is the case, the only new procedures from
this page that you need to type in are: CP16 and DRAW.MANY.MOTIONS.

Given a point and a positive number R, construct the circle of radius R about the point

This page uses concepts from I17, and is a building block for the remainder of the programs.

Instructions: To activate this set of procedures, type CP17 followed by the coordinates of the center, followed by the radius. You must enter the center in brackets, not parentheses, with the coordinates separated by a space, not a comma.

Example: To construct the circle of radius 6 about the center point (-3,2), type: CP17 [-3 2] 6
   The computer draws this:

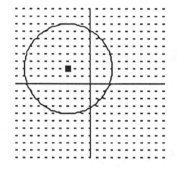

All of the procedures needed to carry out the above functions are listed below. They must be typed into the computer exactly as shown. Remember: any line that is indented from the margin is a continuation of the previous line.

```
TO CP17 :CNTR :RAD
TURNOFFWRAP
SPLIT CLEARTEXT
DRAWCOORDS
MAKE "CNTR ADJUSTPOINT :CNTR
MAKE "RAD 6 * :RAD
DRAW.CIRCLE :CNTR :RAD
END

TO DRAW.CIRCLE :CNTR :RAD
DRAWPT :CNTR
PU SETPOS :CNTR
SETH Ø FD :RAD
MAKE "OLDPT POS
REPEAT 12Ø [MAP.CIRCLE.POINT :CNTR ·RAD]
END

TO MAP.CIRCLE.POINT :CNTR :RAD
PU SETPOS :CNTR
RT 3
FD :RAD
MAKE "NEWPT POS
DRAWSEG :OLDPT :NEWPT
MAKE "OLDPT :NEWPT
END
```

```
TO TURNOFFWRAP
HT WINDOW
END

TO SPLIT
SS
END

TO DRAWCOORDS
HOME CS
DRAWGRID
END

TO DRAWGRID
PD REPEAT 4 [FD 6Ø BK 6Ø RT 9Ø] PU
REPEAT 2 [RT 9Ø FD 6Ø] RT 9Ø
REPEAT 21 [REPEAT 21 [PD FD 1 PU FD 5] BK 126 RT 9Ø FD 6 LT 9Ø]
PU HOME
END

TO DRAWPT :PT
PU SETPOS :PT PD
MAKE "OLDHEAD HEADING
CENTERSQ 1
CENTERSQ 2
SETH :OLDHEAD
END

TO CENTERSQ :RADIUS
PU FD :RADIUS RT 9Ø
PD REPEAT 4 [FD :RADIUS RT 9Ø FD :RADIUS]
PU LT 9Ø BK :RADIUS
END

TO DRAWSEG :PT1 :PT2
PU SETPOS :PT1
PD SETPOS :PT2
END

TO ADJUSTPOINT :P
OP SPROD 6 :P
END

TO SPROD :SCALAR :PAIR
OP LIST :SCALAR * (FIRST :PAIR) :SCALAR * (LAST :PAIR)
END
```

The purpose of each of the above procedures is summarized below, if
it has not been summarized on a previous page.

CP17 first activates TURNOFFWRAP, clears all text, and activates SPLIT
    plus DRAWCOORDS. It then runs the center through ADJUSTPOINT and
    multiplies the radius by 6, to fit the grid, before moving on to
    DRAW.CIRCLE.
DRAW.CIRCLE begins by drawing the center point and placing the turtle
    there. It is set to a heading of 0°, and sent out to the edge of
    the circle with FD :RAD. The position is stored in :OLDPT, and
    then MAP.CIRCLE.POINT is repeated 120 times, since that procedure

draws a piece of the circle whose central angle is three degrees (and 120x3˚=360˚).

MAP.CIRCLE.POINT first places the turtle at the center point, and turns it right 3˚. The turtle goes out to the edge of the circle, and its current position is stored in :NEWPT. A segment is drawn between :OLDPT and :NEWPT, and then the :NEWPT from this run-through becomes the :OLDPT for the next run-through. Notice that the "circle" is actually a 120-sided polygon, but the sides are so small that one can't tell the difference between this and a true circle.

<u>Please note</u>: This CP-page *does* stand alone, and all of the procedures necessary to carry out its purpose are listed directly after the example at the beginning. However, you may notice that many of the above procedures have appeared previously. *If* it so happens that you have already defined (typed in or loaded from disk) all of the procedures from the CP2 page since you started up LOGO, then LOGO already knows how to perform some of the above procedures. If this is the case, the only new procedures from this page that you need to type in are: CP17, DRAW.CIRCLE, and MAP.CIRCLE.POINT.

Given three points in the plane, construct the unique circle that passes through all three points

This page uses concepts from C17 and C17e.

Instructions: To activate this set of procedures, type CP18
followed by the coordinates of the three points. You must enter the
center in brackets, not parentheses, with the coordinates separated
by a space, not a comma. Notice that there is *only* a unique circle
that passes through three distinct points if the three points do not
lie on the same line. If you enter points that lie along a line,
the program should warn you and stop.

Example: To construct the unique circle that passes through the
points (-4,-3), (1,-8), and (8,-5), type: CP18 [-4 -3] [1 -8] [8 -5]
    The computer will draw:

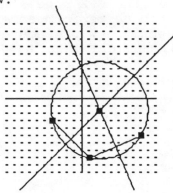

All of the procedures needed to carry out the above functions are
listed below. They must be typed into the computer exactly as
shown. Remember: any line that is indented from the margin is a
continuation of the previous line.

```
TO CP18 :PT1 :PT2 :PT3
IF AND ((LAST :PT1) = (LAST :PT2)) ((LAST :PT2) = (LAST :PT3)) [PR [THE POINTS
    MAY NOT LIE ON THE SAME LINE.] STOP]
TURNOFFWRAP
MAKE "CNTR FIND.CENTER :PT1 :PT2 :PT3
IF :CNTR = "NOCENTER [PR [THE POINTS MAY NOT LIE ON THE SAME LINE.] STOP]
SPLIT CLEARTEXT
DRAWCOORDS
DRAW.CONSTRUCTION :PT1 :PT2 :PT3
DRAW.CIRCLE ADJUSTPOINT :CNTR FDIST ADJUSTPOINT :CNTR ADJUSTPOINT :PT1
END
```

```
TO FIND.CENTER :PT1 :PT2 :PT3
LOCAL "TEMP
IF (LAST :PT1) = (LAST :PT2) [MAKE "TEMP :PT2 MAKE "PT2 :PT3 MAKE "PT3 :TEMP]
IF (LAST :PT2) = (LAST :PT3) [MAKE "TEMP :PT2 MAKE "PT2 :PT1 MAKE "PT1 :TEMP]
MAKE "MID12 MIDPOINT :PT1 :PT2
MAKE "MID23 MIDPOINT :PT2 :PT3
MAKE "SLOPE12PRP (LAST PERP VECTOR :PT1 :PT2) / (FIRST PERP VECTOR :PT1 :PT2)
MAKE "SLOPE23PRP (LAST PERP VECTOR :PT2 :PT3) / (FIRST PERP VECTOR :PT2 :PT3)
IF Ø = ROUND (1ØØ * (:SLOPE12PRP - :SLOPE23PRP)) [OP "NOCENTER]
MAKE "YINT12P (LAST :MID12) + (((FIRST :MID12) * -1) * :SLOPE12PRP)
MAKE "YINT23P (LAST :MID23) + (((FIRST :MID23) * -1) * :SLOPE23PRP)
OP CRAMERS.RULE FIRST :PT1 LAST :PT1 FIRST :PT2 LAST :PT2 FIRST :PT3 LAST :PT3
END


TO CRAMERS.RULE :X1 :Y1 :X2 :Y2 :X3 :Y3
LOCAL "CRAMERD LOCAL "CRAMERX LOCAL "CRAMERY
MAKE "CRAMERD ((:X2 - :X1) * (:Y3 - :Y2)) - ((:X3 - :X2) * (:Y2 - :Y1))
MAKE "CRAMERX ((:Y2 - :Y1) * (:Y3 - :Y2) * :YINT12P) - ((:Y3 - :Y2) * (:Y2 -
   :Y1) * :YINT23P)
MAKE "CRAMERY ((:X2 - :X1) * (:Y3 - :Y2) * :YINT23P) - ((:X3 - :X2) * (:Y2 -
   :Y1) * :YINT12P)
OP LIST (:CRAMERX / :CRAMERD) (:CRAMERY / :CRAMERD)
END


TO DRAW.CONSTRUCTION :PT1 :PT2 :PT3
DRAW.BISECTOR ADJUSTPOINT :PT1 ADJUSTPOINT :PT2
DRAW.BISECTOR ADJUSTPOINT :PT2 ADJUSTPOINT :PT3
DRAWPT ADJUSTPOINT :CNTR
END


TO DRAW.BISECTOR :PT1 :PT2
DRAWPT :PT1
DRAWPT :PT2
DRAWSEG :PT1 :PT2
DRAWLINE MIDPOINT :PT1 :PT2 SUM.POINT.VECT MIDPOINT :PT1 :PT2 PERP VECTOR :PT1
   :PT2
END


TO DRAW.CIRCLE :CNTR :RAD
DRAWPT :CNTR
PU SETPOS :CNTR
SETH Ø FD :RAD
MAKE "OLDPT POS
REPEAT 12Ø [MAP.CIRCLE.POINT :CNTR :RAD]
END


TO MAP.CIRCLE.POINT :CNTR :RAD
PU SETPOS :CNTR
RT 3
FD :RAD
MAKE "NEWPT POS
DRAWSEG :OLDPT :NEWPT
MAKE "OLDPT :NEWPT
END


TO DRAWPT :PT
PU SETPOS :PT PD
MAKE "OLDHEAD HEADING
CENTERSQ 1
CENTERSQ 2
SETH :OLDHEAD
END
```

```
TO CENTERSQ :RADIUS
PU FD :RADIUS RT 9Ø
PD REPEAT 4 [FD :RADIUS RT 9Ø FD :RADIUS]
PU LT 9Ø BK :RADIUS
END


TO DRAWSEG :PT1 :PT2
PU SETPOS :PT1
PD SETPOS :PT2
END


TO TURNOFFWRAP
HT WINDOW
END


TO SPLIT
SS
END


TO DRAWCOORDS
HOME CS
DRAWGRID
END


TO DRAWGRID
PD REPEAT 4 [FD 6Ø BK 6Ø RT 9Ø] PU
REPEAT 2 [RT 9Ø FD 6Ø] RT 9Ø
REPEAT 21 [REPEAT 21 [PD FD 1 PU FD 5] BK 126 RT 9Ø FD 6 LT 9Ø]
PU HOME
END


TO FDIST :PTX :PTY
OP HYP (FIRST :PTX) - (FIRST :PTY) (LAST :PTX) - (LAST :PTY)
END


TO HYP :SIDEA :SIDEB
OP SQRT (:SIDEA * :SIDEA) + (:SIDEB * :SIDEB)
END


TO ADJUSTPOINT :P
OP SPROD 6 :P
END


TO SPROD :SCALAR :PAIR
OP LIST :SCALAR * (FIRST :PAIR) :SCALAR * (LAST :PAIR)
END


TO VECTOR :P :Q
OP LIST (FIRST :Q) - (FIRST :P) (LAST :Q) - (LAST :P)
END


TO PERP :VECT
OP LIST -1 * LAST :VECT FIRST :VECT
END


TO SUM.POINT.VECT :P :Q
OP LIST (FIRST :P) + (FIRST :Q) (LAST :P) + (LAST :Q)
END


TO MIDPOINT :PT1 :PT2
OP LIST MIDCOORD FIRST :PT1 FIRST :PT2 MIDCOORD LAST :PT1 LAST :PT2
END
```

```
TO MIDCOORD :P :Q
OP (:P + :Q) / 2
END


TO DRAWLINE :PT1 :PT2
DRAWSEG RAYEND :PT2 :PT1 RAYEND :PT1 :PT2
END


TO RAYEND :PT1 :PT2
OP FINDRAYEND :PT1 :PT2 16Ø Ø
END


TO FINDRAYEND :PT1 :PT2 :MAXDIST :PARAM
LOCAL "NEWRAYEND
MAKE "NEWRAYEND PARAMLINEPT :PT1 :PT2 :PARAM
IF 16Ø < FDIST [Ø Ø] :NEWRAYEND [OP :NEWRAYEND]
OP FINDRAYEND :PT1 :PT2 :MAXDIST :PARAM + 1
END


TO PARAMLINEPT :PT1 :PT2 :PARAM
OP LIST PARAMNUM FIRST :PT1 FIRST :PT2 :PARAM PARAMNUM LAST :PT1 LAST :PT2
    :PARAM
END


TO PARAMNUM :NUM1 :NUM2 :PARAM
OP :NUM1 + ((:NUM2 - :NUM1) * :PARAM)
END
```

The purpose of each of the above procedures is summarized below, if it has not been summarized on a previous page.

CP18 first checks to see if the three input points all have the same y-coordinate. If so, the slopes of the perpendicular bisectors will all be undefined and will cause an error in FIND.CENTER. TURNOFFWRAP is activated, and the output of FIND.CENTER for the three entered points is stored as the :CNTR of the circle. If FIND.CENTER has output "NOCENTER it means that the perpendicular bisectors never intersect, and the points must all lie on a line. It then activates SPLIT, clears all text, and activates DRAWCOORDS and DRAW.CONSTRUCTION. It finally draws the determined circle by using DRAW.CIRCLE to draw a circle with center :CNTR and a radius of the distance from :CNTR to the first point entered (it could have been any of them, since they're all the same distance from :CNTR).

FIND.CENTER and CRAMERS.RULE work together to find the intersection point of the perpendicular bisectors of the two segments between the first point and the second point, and the second point and the third point respectively. The point of intersection is found using *Cramer's Rule*, which is too complicated to explain here but can be found in many second-year algebra texts. Suffice it to say that it requires the coordinates of the three points and the determined y-intercepts of the perpendicular bisectors of the segments mentioned above. FIND.CENTER finds the y-intercepts, and CRAMERS.RULE uses the rule to output the final point of intersection. FIND.CENTER begins by making sure that no two adjacent points in the series of three points have y-coordinates that are identical. If they did, the slope of their perpendicular bisector would be undefined and FIND.CENTER wouldn't

work. So, if the first and second entered points have the same
y-coordinate, the program swaps the third point and the second
point. If the second and third have identical y-coordinates, the
first and second are swapped. In this way, the slopes of the
perpendicular bisectors found later are never undefined. Next,
the midpoints of the segments are found, and the slopes of the
perpendicular bisectors found. At this point the procedure stops
and outputs "NOCENTER to :CNTR if the slopes are the same or very
close, because if they are they will never intersect (meaning
that the three entered points all lie on the same line and no
circle exists). Next the y-intercepts are found from the
traditional y = mx + b equation of the bisectors, and thus
CRAMERS.RULE can do its job.

DRAW.CONSTRUCTION draws the perpendicular bisectors plus the
determined center point, and in the process draws the three
entered points.

<u>Please note</u>: This CP-page *does* stand alone, and all of the
procedures necessary to carry out its purpose are listed directly
after the example at the beginning. However, you may notice that
many of the above procedures have appeared previously. *If* it so
happens that you have already defined (typed in or loaded from disk)
all of the procedures from the CP2 page, the CP3 page, the CP4 page,
and the CP17 page since you started up LOGO, then LOGO already knows
how to perform most of the above procedures. If this is the case,
the only new procedures from this page that you need to type in are:
CP18, FIND.CENTER, CRAMERS.RULE and DRAW.CONSTRUCTION.

Given the center of a circle and a point on the circle, construct the tangent to the circle through the point

This page uses concepts from I25.

Instructions: To activate this set of procedures, type CP19 followed by the coordinates of the center of the circle, followed by the coordinates of the point on the circle. The radius will be figured automatically. You must enter the points in brackets, not parentheses, with the coordinates separated by a space, not a comma.

Example: To construct the tangent to the circle of center (-2,-3) that passes through the point (3,1) on the circle, type:
CP19 [-2 -3] [3 1]
     The computer draws:

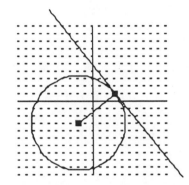

All of the procedures needed to carry out the above functions are listed below. They must be typed into the computer exactly as shown. Remember: any line that is indented from the margin is a continuation of the previous line.


```
TO CP19 :CNTR :PT
TURNOFFWRAP
SPLIT CLEARTEXT
DRAWCOORDS
MAKE "PT ADJUSTPOINT :PT
MAKE "CNTR ADJUSTPOINT :CNTR
DRAW.TANGENT :CNTR :PT
END

TO DRAW.TANGENT :CNTR :PT
DRAW.CIRCLE :CNTR FDIST :CNTR :PT
DRAWPT :PT
DRAWSEG :CNTR :PT
DRAWLINE :PT SUM.POINT.VECT :PT PERP VECTOR :CNTR :PT
END

TO DRAW.CIRCLE :CNTR :RAD
DRAWPT :CNTR
PU SETPOS :CNTR
SETH Ø FD :RAD
MAKE "OLDPT POS
REPEAT 12Ø [MAP.CIRCLE.POINT :CNTR :RAD]
END
```

```
TO MAP.CIRCLE.POINT :CNTR :RAD
PU SETPOS :CNTR
RT 3
FD :RAD
MAKE "NEWPT POS
DRAWSEG :OLDPT :NEWPT
MAKE "OLDPT :NEWPT
END


TO DRAWPT :PT
PU SETPOS :PT PD
MAKE "OLDHEAD HEADING
CENTERSQ 1
CENTERSQ 2
SETH :OLDHEAD
END


TO CENTERSQ :RADIUS
PU FD :RADIUS RT 90
PD REPEAT 4 [FD :RADIUS RT 90 FD :RADIUS]
PU LT 90 BK :RADIUS
END


TO DRAWSEG :PT1 :PT2
PU SETPOS :PT1
PD SETPOS :PT2
END


TO TURNOFFWRAP
HT WINDOW
END


TO SPLIT
SS
END


TO DRAWCOORDS
HOME CS
DRAWGRID
END


TO DRAWGRID
PD REPEAT 4 [FD 60 BK 60 RT 90] PU
REPEAT 2 [RT 90 FD 60] RT 90
REPEAT 21 [REPEAT 21 [PD FD 1 PU FD 5] BK 126 RT 90 FD 6 LT 90]
PU HOME
END


TO FDIST :PTX :PTY
OP HYP (FIRST :PTX) - (FIRST :PTY) (LAST :PTX) - (LAST :PTY)
END


TO HYP :SIDEA :SIDEB
OP SQRT (:SIDEA * :SIDEA) +  (:SIDEB * :SIDEB)
END


TO ADJUSTPOINT :P
OP SPROD 6 :P
END


TO SPROD :SCALAR :PAIR
OP LIST :SCALAR * (FIRST :PAIR) :SCALAR * (LAST :PAIR)
END
```

```
TO VECTOR :P :Q
OP LIST (FIRST :Q) - (FIRST :P) (LAST :Q) - (LAST :P)
END


TO PERP :VECT
OP LIST -1 * LAST :VECT FIRST :VECT
END


TO SUM.POINT.VECT :P :Q
OP LIST (FIRST :P) + (FIRST :Q) (LAST :P) + (LAST :Q)
END


TO DRAWLINE :PT1 :PT2
DRAWSEG RAYEND :PT2 :PT1 RAYEND :PT1 :PT2
END


TO RAYEND :PT1 :PT2
OP FINDRAYEND :PT1 :PT2 16Ø Ø
END


TO FINDRAYEND :PT1 :PT2 :MAXDIST :PARAM
LOCAL "NEWRAYEND
MAKE "NEWRAYEND PARAMLINEPT :PT1 :PT2 :PARAM
IF 16Ø < FDIST [Ø Ø] :NEWRAYEND [OP :NEWRAYEND]
OP FINDRAYEND :PT1 :PT2 :MAXDIST :PARAM + 1
END


TO PARAMLINEPT :PT1 :PT2 :PARAM
OP LIST PARAMNUM FIRST :PT1 FIRST :PT2 :PARAM PARAMNUM LAST :PT1 LAST :PT2
   :PARAM
END


TO PARAMNUM :NUM1 :NUM2 :PARAM
OP :NUM1 + ((:NUM2 - :NUM1) * :PARAM)
END
```

The purpose of each of the above procedures is summarized below, if
it has not been summarized on a previous page.

CP19 first activates TURNOFFWRAP, clears all text, and activates SPLIT
   plus DRAWCOORDS. It then moves directly to DRAW.TANGENT.
DRAW.TANGENT begins by running the two entered points through
   ADJUSTPOINT. It then draws the circle of center :CNTR and with a
   radius of the distance from :CNTR to :PT. It then draws the point
   and the radius of the circle from :CNTR to :PT. Last, it draws
   the tangent by drawing the line that passes through 1) the point
   on the circle and 2) a point whose difference from the entered
   point is a vector perpendicular to the vector from :CNTR to :PT.

Please note: This CP-page *does* stand alone, and all of the
procedures necessary to carry out its purpose are listed directly
after the example at the beginning. However, you may notice that
many of the above procedures have appeared previously. *If* it so
happens that you have already defined (typed in or loaded from disk)
all of the procedures from the CP2 page, the CP3 page, and the CP17

page since you started up LOGO, then LOGO already knows how to
perform most of the above procedures. If this is the case, the only
new procedures from this page that you need to type in are: CP19 and
DRAW.TANGENT.

Given a circle and a point outside the circle, construct the two lines tangent to the circle that pass through the point

This page uses concepts from I24e, I25, and I25e.

<u>Instructions</u>: To activate this set of procedures, type CP20 followed by the coordinates of the center of the circle, followed by the radius of the circle, followed by the coordinates of the point outside the circle. You must enter the center in brackets, not parentheses, with the coordinates separated by a space, not a comma. If your point is not outside the circle, the program will give a warning and stop. Since it takes a while for the computer to calculate each tangent line, allow it some time to work.

<u>Example</u>: To construct the two tangent lines to the circle of radius 6 about center (-2,-3) that pass through the point (7,6) outside the circle, type: CP20 [-2 -3] 6 [7 6]
    The computer draws this:

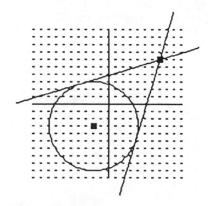

All of the procedures needed to carry out the above functions are listed below. They must be typed into the computer exactly as shown. Remember: any line that is indented from the margin is a continuation of the previous line.

```
TO CP2Ø :CNTR :RAD :PT
TURNOFFWRAP
IF (FDIST :CNTR :PT) < (:RAD + 1)   [PR [YOUR POINT IS INSIDE OR ON CIRCLE!]
    STOP]
SPLIT CLEARTEXT
DRAWCOORDS
MAKE "CNTR ADJUSTPOINT :CNTR
MAKE "PT ADJUSTPOINT :PT
MAKE "RAD 6 * :RAD
DRAW.CIRCLE :CNTR :RAD
DRAW.BOTH.TANGENTS :CNTR :RAD :PT
END
```

```
TO DRAW.BOTH.TANGENTS :CNTR :RAD :PT
MAKE "CNTRMID MIDPOINT :CNTR :PT
MAKE "RADMID FDIST :CNTRMID :CNTR
SPLIT CLEARTEXT PR [*** CALCULATING TANGENTS ***]
PR [PLEASE STAND BY...]
DRAWPT :PT
DRAWLINE :PT FIND.TAN.PT :CNTR :RAD :CNTRMID :RADMID 1
DRAWLINE :PT FIND.TAN.PT :CNTR :RAD :CNTRMID :RADMID -1
CLEARTEXT
END


TO FIND.TAN.PT :CNTR :RAD :CNTRMID :RADMID :STEP
PU SETPOS :CNTR SETH HEADTOWARDS :CNTRMID
MAKE "LEASTDIF 100000
REPEAT 93 [MAKE "LEASTDIF FIND.LEAST.DIF :RAD :CNTRMID :RADMID :STEP]
PU SETPOS :CNTR SETH HEADTOWARDS :CNTRMID
REPEAT 93 [FIND.TANPT.POS :LEASTDIF :RAD :CNTRMID :RADMID :STEP]
OP :FOUNDTANPT
END


TO FIND.LEAST.DIF :RAD :CNTRMID :RADMID :STEP
LOCAL "DISTCNTRMID LOCAL "DIFDIST
FD :RAD
MAKE "DISTCNTRMID FDIST :CNTRMID POS
MAKE "DIFDIST ABSVAL (:RADMID - :DISTCNTRMID)
IF :DIFDIST < :LEASTDIF [MAKE "LEASTDIF :DIFDIST]
BK :RAD RT :STEP
OP :LEASTDIF
END


TO FIND.TANPT.POS :LEASTDIF :RAD :CNTRMID :RADMID :STEP
LOCAL "DISTCNTRMID LOCAL "DIFDIST
FD :RAD
MAKE "DISTCNTRMID FDIST :CNTRMID POS
MAKE "DIFDIST ABSVAL (:RADMID - :DISTCNTRMID)
IF :DIFDIST = :LEASTDIF [MAKE "FOUNDTANPT POS]
BK :RAD RT :STEP
END


TO ABSVAL :NUM
IF :NUM < 0 [OP -1 * :NUM]
OP :NUM
END


TO HEADTOWARDS :PAIR
OP TOWARDS :PAIR
END


TO DRAW.CIRCLE :CNTR :RAD
DRAWPT :CNTR
PU SETPOS :CNTR
SETH 0 FD :RAD
MAKE "OLDPT POS
REPEAT 120 [MAP.CIRCLE.POINT :CNTR :RAD]
END
```

```
TO MAP.CIRCLE.POINT :CNTR :RAD
PU SETPOS :CNTR
RT 3
FD :RAD
MAKE "NEWPT POS
DRAWSEG :OLDPT :NEWPT
MAKE "OLDPT :NEWPT
END

TO DRAWPT :PT
PU SETPOS :PT PD
MAKE "OLDHEAD HEADING
CENTERSQ 1
CENTERSQ 2
SETH :OLDHEAD
END

TO CENTERSQ :RADIUS
PU FD :RADIUS RT 9Ø
PD REPEAT 4 [FD :RADIUS RT 9Ø FD :RADIUS]
PU LT 9Ø BK :RADIUS
END

TO DRAWSEG :PT1 :PT2
PU SETPOS :PT1
PD SETPOS :PT2
END

TO TURNOFFWRAP
HT WINDOW
END

TO SPLIT
SS
END

TO DRAWCOORDS
HOME CS
DRAWGRID
END

TO DRAWGRID
PD REPEAT 4 [FD 6Ø BK 6Ø RT 9Ø] PU
REPEAT 2 [RT 9Ø FD 6Ø] RT 9Ø
REPEAT 21 [REPEAT 21 [PD FD 1 PU FD 5] BK 126 RT 9Ø FD 6 LT 9Ø]
PU HOME
END

TO FDIST :PTX :PTY
OP HYP (FIRST :PTX) - (FIRST :PTY) (LAST :PTX) - (LAST :PTY)
END

TO HYP :SIDEA :SIDEB
OP SQRT (:SIDEA * :SIDEA) +  (:SIDEB * :SIDEB)
END

TO ADJUSTPOINT :P
OP SPROD 6 :P
END

TO SPROD :SCALAR :PAIR
OP LIST :SCALAR * (FIRST :PAIR) :SCALAR * (LAST :PAIR)
END
```

```
TO MIDPOINT :PT1 :PT2
OP LIST MIDCOORD FIRST :PT1 FIRST :PT2 MIDCOORD LAST :PT1 LAST :PT2
END


TO MIDCOORD :P :Q
OP (:P + :Q) / 2
END


TO DRAWLINE :PT1 :PT2
DRAWSEG RAYEND :PT2 :PT1 RAYEND :PT1 :PT2
END


TO RAYEND :PT1 :PT2
OP FINDRAYEND :PT1 :PT2 16Ø Ø
END


TO FINDRAYEND :PT1 :PT2 :MAXDIST :PARAM
LOCAL "NEWRAYEND
MAKE "NEWRAYEND PARAMLINEPT :PT1 :PT2 :PARAM
IF 16Ø < FDIST [Ø Ø] :NEWRAYEND [OP :NEWRAYEND]
OP FINDRAYEND :PT1 :PT2 :MAXDIST :PARAM + 1
END


TO PARAMLINEPT :PT1 :PT2 :PARAM
OP LIST PARAMNUM FIRST :PT1 FIRST :PT2 :PARAM PARAMNUM LAST :PT1 LAST :PT2
   :PARAM
END


TO PARAMNUM :NUM1 :NUM2 :PARAM
OP :NUM1 + ((:NUM2 - :NUM1) * :PARAM)
END
```

The purpose of each of the above procedures is summarized below, if
it has not been summarized on a previous page.

CP20 first activates TURNOFFWRAP, and checks to make sure that the
   point entered is outside the circle. It then clears all text,
   and activates SPLIT plus DRAWCOORDS. It then runs the center and
   point through ADJUSTPOINT and multiplies the radius by 6, to fit
   the grid, before moving on to DRAW.CIRCLE and DRAW.BOTH.TANGENTS.
DRAW.BOTH.TANGENTS uses the method from exercise #8 on I25e to find
   the two tangent lines. It creates the circle whose center is the
   midpoint of the segment from the input center to the input point,
   and whose radius is the distance from this midpoint to the input
   center. Where this new circle intersects the original circle is
   where the two tangent lines must pass. The intersection points
   of these two circles are discovered in FIND.TAN.PT. This
   procedure begins by storing the center of the aforementioned
   circle in :CNTRMID and the radius in :RADMID. After printing
   instructions it draws the input point and draws a line through
   this point and each tangent point found from each activation of
   FIND.TAN.PT.
FIND.TAN.PT finds a tangent point by finding the intersection of the
   two circles discussed above. It does this by finding a place
   where the turtle's distance from the input center is the input
   radius and at the same time its distance from the :CNTRMID is very
   close to the :RADMID. The turtle is sent out to the edge of the
   input circle in 93 increments of one degree. When it is at the

edge of the input circle, its distance from the input center is automatically the input radius, so all that needs to be checked is how its distance from :CNTRMID compares to :RADMID. Each time, the turtle's distance from :CNTRMID is stored in :DISTCNTRMID, and the difference between :DISTCNTRMID and :RADMID is stored in :LEASTDIF. The object is to find out where the smallest :LEASTDIF occurs; this is where the turtle's distance from :CNTRMID is closest to :RADMID, so it is where the tangent point lies. This procedure starts by placing the turtle on the input center and setting its heading toward :CNTRMID. The :LEASTDIF variable is then set to some very large number, so that it can be replaced by an appropriate :DIFDIST, later. First, FIND.LEAST.DIF in activated 93 times to find out what the miniumum possible :LEASTDIF is, and then FIND.TANPT.POS send the turtle through the same route, recording a tangent point if the turtle's position gives that minimum :LEASTDIF. That tangent point, once found, is output back to DRAW.BOTH.TANGENTS.

FIND.LEAST.DIF begins by sending the turtle out to the edge of the original circle so that its distance from the input center is the input radius. Its distance from :CNTRMID is stored in :DISTCNTRMID, and the absolute value of the difference between :RADMID and :DISTCNTRMID becomes :DIFDIST. If this :DIFDIST is even smaller than the last :LEASTDIF that was found, it becomes the new :LEASTDIF. So, the lowest possible :DIFDIST can be found in :LEASTDIF when the procedure's 93 activations, one for each degree of one side of the original circle that faces the input point, are complete. At the end of this procedure, the turtle goes back to the input center and turns right either 1˚ or -1˚ (left 1˚) depending on the :STEP from DRAW.BOTH.TANGENTS. In this way, the entire side of the original circle that faces the input point can be covered in the tangent point search.

FIND.TANPT.POS also begins by sending the turtle out to the edge of the original circle. The :DIFDIST for its current position is found just as above, and if this :DIFDIST is the same as the :LEASTDIF that was found above, then the turtle's current position is where the tangent line must pass. This becomes :FOUNDTANPT.

ABSVAL outputs the absolute value of the input (i.e. the input if the input was positive, or the input with the opposite sign if the input was negative).

HEADTOWARDS outputs the heading that the turtle would need to take in order to go directly to the input point.

<u>Please note</u>: This CP-page *does* stand alone, and all of the procedures necessary to carry out its purpose are listed directly after the example at the beginning. However, you may notice that many of the above procedures have appeared previously. *If* it so happens that you have already defined (typed in or loaded from disk) all of the procedures from the CP2 page, the CP4 page, and the CP17 page since you started up LOGO, then LOGO already knows how to perform many of the above procedures. If this is the case, the only new procedures from this page that you need to type in are: CP20, DRAW.BOTH.TANGENTS, FIND.TAN.PT, FIND.LEAST.DIF, FIND.TANPT.POS, ABSVAL, and HEADTOWARDS.

Given a point X inside or outside the circle of radius one and center O, construct the reciprocal point X'

This page uses concepts from C23.

Instructions: To activate this set of procedures, type CP21, followed by the coordinates of X. You must enter the point in brackets, not parentheses, with the coordinates separated by a space, not a comma. Note that this grid, to show the unit circle, is on a very small scale, so neither your x or y coordinate should be much larger than 3.

Example: To construct the reciprocal point of (0.3,0.5), type:
CP21 [0.3 0.5]
    The computer draws this:

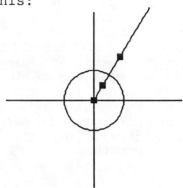

All of the procedures needed to carry out the above functions are listed below. They must be typed into the computer exactly as shown. Remember: any line that is indented from the margin is a continuation of the previous line.

```
TO CP21 :PT
TURNOFFWRAP
SPLIT CLEARTEXT
HOME CS
DRAW.AXES
DRAW.CIRCLE [Ø Ø] 3Ø
DRAW.RECIPROCAL :PT
END

TO DRAW.RECIPROCAL :PT
DRAWPT FIT.HYP.GRID :PT
DRAWPT FIT.HYP.GRID RECIPROCAL.PT :PT
DRAWRAY [Ø Ø] FIT.HYP.GRID RECIPROCAL.PT :PT
END

TO FIT.HYP.GRID :PT
OP SPROD 3Ø :PT
END
```

```
TO DRAW.AXES
PD REPEAT 4 [FD 7Ø BK 7Ø RT 9Ø] PU
END


TO RECIPROCAL.PT :X
MAKE "OX FDIST [Ø Ø] :X
MAKE "OX1 1 / :OX
OP SPROD (:OX1 / :OX) :X
END


TO DRAW.CIRCLE :CNTR :RAD
DRAWPT :CNTR
PU SETPOS :CNTR
SETH Ø FD :RAD
MAKE "OLDPT POS
REPEAT 12Ø [MAP.CIRCLE.POINT :CNTR :RAD]
END


TO MAP.CIRCLE.POINT :CNTR :RAD
PU SETPOS :CNTR
RT 3
FD :RAD
MAKE "NEWPT POS
DRAWSEG :OLDPT :NEWPT
MAKE "OLDPT :NEWPT
END


TO DRAWPT :PT
PU SETPOS :PT PD
MAKE "OLDHEAD HEADING
CENTERSQ 1
CENTERSQ 2
SETH :OLDHEAD
END


TO CENTERSQ :RADIUS
PU FD :RADIUS RT 9Ø
PD REPEAT 4 [FD :RADIUS RT 9Ø FD :RADIUS]
PU LT 9Ø BK :RADIUS
END


TO DRAWSEG :PT1 :PT2
PU SETPOS :PT1
PD SETPOS :PT2
END


TO TURNOFFWRAP
HT WINDOW
END


TO SPLIT
SS
END


TO DRAWRAY :PT1 :PT2
DRAWSEG :PT1 RAYEND :PT1 :PT2
END


TO RAYEND :PT1 :PT2
OP FINDRAYEND :PT1 :PT2 16Ø Ø
END
```

```
TO FINDRAYEND :PT1 :PT2 :MAXDIST :PARAM
LOCAL "NEWRAYEND
MAKE "NEWRAYEND PARAMLINEPT :PT1 :PT2 :PARAM
IF 16Ø < FDIST [Ø Ø] :NEWRAYEND [OP :NEWRAYEND]
OP FINDRAYEND :PT1 :PT2 :MAXDIST :PARAM + 1
END


TO PARAMLINEPT :PT1 :PT2 :PARAM
OP LIST PARAMNUM FIRST :PT1 FIRST :PT2 :PARAM PARAMNUM LAST :PT1 LAST :PT2
    :PARAM
END


TO PARAMNUM :NUM1 :NUM2 :PARAM
OP :NUM1 + ((:NUM2 - :NUM1) * :PARAM)
END


TO FDIST :PTX :PTY
OP HYP (FIRST :PTX) - (FIRST :PTY) (LAST :PTX) - (LAST :PTY)
END


TO HYP :SIDEA :SIDEB
OP SQRT (:SIDEA * :SIDEA) +  (:SIDEB * :SIDEB)
END


TO SPROD :SCALAR :PAIR
OP LIST :SCALAR * (FIRST :PAIR) :SCALAR * (LAST :PAIR)
END
```

The purpose of each of the above procedures is summarized below, if
it has not been summarized on a previous page.

CP21 first activates TURNOFFWRAP, clears all text and graphics, and
    activates SPLIT plus DRAW.AXES. It then draws the unit circle, and
    activates DRAW.RECIPROCAL.
DRAW.RECIPROCAL draws the entered point after running it through
    FIT.HYP.GRID, and does the same for the RECIPROCAL.PT of the point.
    It then draws the ray that contains both points.
FIT.HYP.GRID outputs the scalar product of the input and 30. This
    greatly magnifies the point so it can be seen appropriately seen
    on the extremely small scale of this grid.
DRAW.AXES draws the x-axis and y-axis.
RECIPROCAL.PT outputs the reciprocal point of its input point X. The
    distance OX is found, and OX' is the inverse of this. In the
    last line, the coordinates of X are each divided by OX to give a
    unit vector in the direction of X, and this is SPROD by OX' to
    give the coordinates of X'. The math is from C23.

Please note: This CP-page *does* stand alone, and all of the
procedures necessary to carry out its purpose are listed directly
after the example at the beginning. However, you may notice that
many of the above procedures have appeared previously. *If* it so
happens that you have already defined (typed in or loaded from disk)
all of the procedures from the CP2 page and the CP17 page since you
started up LOGO, then LOGO already knows how to perform many of the
above procedures. If this is the case, the only new procedures from
this page that you need to type in are: CP21, DRAW.RECIPROCAL,
FIT.HYP.GRID, DRAW.AXES, and RECIPROCAL.PT.

Given two points inside the circle of radius one about (0,0), construct the hyperbolic line containing the two points

This page uses concepts from I40 and C23.

Instructions: To activate this set of procedures, type CP22, followed by the coordinates of each of the two points. You must enter the points in brackets, not parentheses, with the coordinates separated by a space, not a comma. Note that the points must lie within the unit circle. If they do not, the program will give a warning and stop.

Example: To construct the hyperbolic line that passes through the two points (0.2,0.4) and (-0.6,0.3), type: CP22 [0.2 0.4] [-0.6 0.3]
    The computer draws:

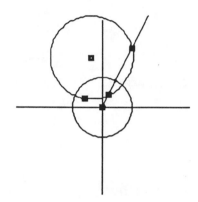

The circle about (0,0) represents hyperbolic-land, and the other circle is the circle one would construct on C23 to obtain the hyperbolic line. Notice how the two circles meet perpendicularly.

All of the procedures needed to carry out the above functions are listed below. They must be typed into the computer exactly as shown. Remember: any line that is indented from the margin is a continuation of the previous line.

```
TO CP22 :PT1 :PT2
TURNOFFWRAP
IF OR ((FDIST [Ø Ø] :PT1) > 1) ((FDIST [Ø Ø] :PT2) > 1) [PR [POINTS MUST BE
    INSIDE UNIT CIRCLE.] STOP]
SPLIT HOME CS
DRAW.AXES
DRAW.CIRCLE [Ø Ø] 3Ø
DRAW.RECIP.AND.POINTS :PT1 :PT2
IF AND ((LAST :PT1) = Ø) ((LAST :PT2) = Ø) [DRAWLINE [Ø Ø] FIT.HYP.GRID :PT1
    STOP]
MAKE "CNTR FIND.CENTER FIT.HYP.GRID :PT1 FIT.HYP.GRID :PT2 FIT.HYP.GRID
    RECIPROCAL.PT :PT1
IF :CNTR = "NOCENTER [DRAWLINE [Ø Ø] FIT.HYP.GRID :PT1 STOP]
DRAW.CIRCLE :CNTR FDIST :CNTR FIT.HYP.GRID :PT1
END
```

```
TO DRAW.RECIP.AND.POINTS :PT1 :PT2
DRAW.RECIPROCAL :PT1
DRAWPT FIT.HYP.GRID :PT2
END


TO FIND.CENTER :PT1 :PT2 :PT3
LOCAL "TEMP
IF (LAST :PT1) = (LAST :PT2) [MAKE "TEMP :PT2 MAKE "PT2 :PT3 MAKE "PT3 :TEMP]
IF (LAST :PT2) = (LAST :PT3) [MAKE "TEMP :PT2 MAKE "PT2 :PT1 MAKE "PT1 :TEMP]
MAKE "MID12 MIDPOINT :PT1 :PT2
MAKE "MID23 MIDPOINT :PT2 :PT3
MAKE "SLOPE12PRP (LAST PERP VECTOR :PT1 :PT2) / (FIRST PERP VECTOR :PT1 :PT2)
MAKE "SLOPE23PRP (LAST PERP VECTOR :PT2 :PT3) / (FIRST PERP VECTOR :PT2 :PT3)
IF Ø = ROUND (100 * (:SLOPE12PRP - :SLOPE23PRP)) [OP "NOCENTER]
MAKE "YINT12P (LAST :MID12) + (((FIRST :MID12) * -1) * :SLOPE12PRP)
MAKE "YINT23P (LAST :MID23) + (((FIRST :MID23) * -1) * :SLOPE23PRP)
OP CRAMERS.RULE FIRST :PT1 LAST :PT1 FIRST :PT2 LAST :PT2 FIRST :PT3 LAST :PT3
END


TO CRAMERS.RULE :X1 :Y1 :X2 :Y2 :X3 :Y3
LOCAL "CRAMERD LOCAL "CRAMERX LOCAL "CRAMERY
MAKE "CRAMERD ((:X2 - :X1) * (:Y3 - :Y2)) - ((:X3 - :X2) * (:Y2 - :Y1))
MAKE "CRAMERX ((:Y2 - :Y1) * (:Y3 - :Y2) * :YINT12P) - ((:Y3 - :Y2) * (:Y2 -
    :Y1) * :YINT23P)
MAKE "CRAMERY ((:X2 - :X1) * (:Y3 - :Y2) * :YINT23P) - ((:X3 - :X2) * (:Y2 -
    :Y1) * :YINT12P)
OP LIST (:CRAMERX / :CRAMERD) (:CRAMERY / :CRAMERD)
END


TO DRAW.CIRCLE :CNTR :RAD
DRAWPT :CNTR
PU SETPOS :CNTR
SETH Ø FD :RAD
MAKE "OLDPT POS
REPEAT 120 [MAP.CIRCLE.POINT :CNTR :RAD]
END


TO MAP.CIRCLE.POINT :CNTR :RAD
PU SETPOS :CNTR
RT 3
FD :RAD
MAKE "NEWPT POS
DRAWSEG :OLDPT :NEWPT
MAKE "OLDPT :NEWPT
END


TO DRAWPT :PT
PU SETPOS :PT PD
MAKE "OLDHEAD HEADING
CENTERSQ 1
CENTERSQ 2
SETH :OLDHEAD
END


TO CENTERSQ :RADIUS
PU FD :RADIUS RT 90
PD REPEAT 4 [FD :RADIUS RT 90 FD :RADIUS]
PU LT 90 BK :RADIUS
END


TO DRAWSEG :PT1 :PT2
PU SETPOS :PT1
PD SETPOS :PT2
END
```

```
TO TURNOFFWRAP
HT WINDOW
END

TO SPLIT
SS
END

TO DRAW.RECIPROCAL :PT
DRAWPT FIT.HYP.GRID :PT
DRAWPT FIT.HYP.GRID RECIPROCAL.PT :PT
DRAWRAY [Ø Ø] FIT.HYP.GRID RECIPROCAL.PT :PT
END

TO RECIPROCAL.PT :X
MAKE "OX FDIST [Ø Ø] :X
MAKE "OX1 1 / :OX
OP SPROD (:OX1 / :OX) :X
END

TO FIT.HYP.GRID :PT
OP SPROD 3Ø :PT
END

TO DRAW.AXES
PD REPEAT 4 [FD 7Ø BK 7Ø RT 9Ø] PU
END

TO DRAWRAY :PT1 :PT2
DRAWSEG :PT1 RAYEND :PT1 :PT2
END

TO DRAWLINE :PT1 :PT2
DRAWSEG RAYEND :PT2 :PT1 RAYEND :PT1 :PT2
END

TO RAYEND :PT1 :PT2
OP FINDRAYEND :PT1 :PT2 16Ø Ø
END

TO FINDRAYEND :PT1 :PT2 :MAXDIST :PARAM
LOCAL "NEWRAYEND
MAKE "NEWRAYEND PARAMLINEPT :PT1 :PT2 :PARAM
IF 16Ø < FDIST [Ø Ø] :NEWRAYEND [OP :NEWRAYEND]
OP FINDRAYEND :PT1 :PT2 :MAXDIST :PARAM + 1
END

TO PARAMLINEPT :PT1 :PT2 :PARAM
OP LIST PARAMNUM FIRST :PT1 FIRST :PT2 :PARAM PARAMNUM LAST :PT1 LAST :PT2
    :PARAM
END

TO PARAMNUM :NUM1 :NUM2 :PARAM
OP :NUM1 + ((:NUM2 - :NUM1) * :PARAM)
END

TO FDIST :PTX :PTY
OP HYP (FIRST :PTX) - (FIRST :PTY) (LAST :PTX) - (LAST :PTY)
END

TO HYP :SIDEA :SIDEB
OP SQRT (:SIDEA * :SIDEA) +  (:SIDEB * :SIDEB)
END
```

```
TO VECTOR :P :Q
OP LIST (FIRST :Q) - (FIRST :P) (LAST :Q) - (LAST :P)
END

TO PERP :VECT
OP LIST -1 * LAST :VECT FIRST :VECT
END

TO MIDPOINT :PT1 :PT2
OP LIST MIDCOORD FIRST :PT1 FIRST :PT2 MIDCOORD LAST :PT1 LAST :PT2
END

TO MIDCOORD :P :Q
OP (:P + :Q) / 2
END

TO SPROD :SCALAR :PAIR
OP LIST :SCALAR * (FIRST :PAIR) :SCALAR * (LAST :PAIR)
END
```

The purpose of each of the above procedures is summarized below, if it has not been summarized on a previous page.

CP22 first activates TURNOFFWRAP, and checks to make sure that the points are inside the unit circle. It then activates SPLIT and clears all graphics before activating DRAW.AXES, drawing the unit circle, and drawing the entered points plus the reciprocal of the first. A preliminary check is made to see if all three points (the two entered and the one reciprocal) all lie along a straight line. If this is the case, the hyperbolic line must pass through (0,0), which means that it appears straight rather than curved, and a straight line is drawn. Then, just as in the C23 construction, the three points are run through FIND.CENTER, which finds the center of the circle that determines the hyperbolic line by finding the intersection of the perpendicular bisectors of segments between the points. If FIND.CENTER outputs "NOCENTER to the :CNTR variable, this means that the hyperbolic line must pass through (0,0) and, just as before, a straight-appearing hyperbolic line is drawn. Lastly, the circle that contains the hyperbolic line is drawn around :CNTR.

DRAW.RECIP.AND.POINTS uses DRAW.RECIPROCAL to draw the entered points, the reciprocal point, and the ray through the reciprocal point.

Please note: This CP-page *does* stand alone, and all of the procedures necessary to carry out its purpose are listed directly after the example at the beginning. However, you may notice that many of the above procedures have appeared previously. *If* it so happens that you have already defined (typed in or loaded from disk) *all* of the procedures from the CP2 page, the CP18 page, and the CP21 page since you started up LOGO, then LOGO already knows how to perform most of the above procedures. If this is the case, the only new procedures from this page that you need to type in are: CP22 and DRAW.RECIP.AND.POINTS.

Index

The index is referenced by section/topic number, rather than by page number. For example, "I31" refers to the Intuition section, topic number 31 (not page 31). Similarly, "C" refers to Construction, "P" refers to Proof, and "CP" refers to Computer Programs.